Recent Progress in Adsorption and Ion Exchange

Y.H. Ma and
J.P. Ausikaitis, editors

R.S. Artigue
Aaron Barkatt
J.M. Begouich
B.F. Brian III
C.H. Brown
P. Burn
D.O. Campbell
E.D. Collins
S.J. Doong
H.L. Fleming
D.C. Grant
Hamid Hojaji
Y.S. Lin
Y.H. Ma
K.A. Michael
G.W. Miller
D.K. Ploetz
S.M. Robinson
A.K. Saha
M.C. Skriba
Ajay Sood
H.G. Sutter
R.T. Yang
Imre Zwiebel

AIChE Symposium Series 259

Number 259 1987 Volume 83

Published by

American Institute of Chemical Engineers

345 East 47 Street New York, New York 10017

Library of Congress Cataloging-in-Publication Data
Recent progress in adsorption and ion exchange / Y.H. Ma and J.P. Ausikaitis, editors.

p. cm. -- (AIChE symposium series ; no. 259, v. 83)
"Contains papers presented at the AIChE 1987 Houston National Meeting"--Foreword.
Includes index.
ISBN 0-8169-0436-7
1. Adsorption—Congresses. 2. Ion exchange—Congresses. I. Ma, Y.H., 1936- II. Ausikaitis, J.P. (Joseph P.), 1949- III. National Meeting of AIChE (1987 : Houston, Tex.) IV. Series: AIChE symposium series ; no. 259.
QD547.R43 1987
541.3'453--dc19 87-31901
CIP

Printed in the United States of America by
Twin Production & Design

FOREWORD

Adsorption has recently gained substantial attention due to its vast potential in separation applications. This volume contains papers presented at the AIChE 1987 Houston National Meeting. The importance of ion exchange processes and adsorption separations in liquid phase is reflected by the large number of papers in these areas being included in this volume. The application of adsorption processes in downstream bioprocessing will be a key element in the development of biotechnology. It is hoped that the present volume will generate more interest in the study of liquid phase adsorption and diffusion.

We wish to thank the authors and reviewers who enabled this volume to be assembled and to the AIChE Publications staff for their patience and efficient and prompt handling of the publication process which makes it possible for the Symposium Series to be published in a relatively short period of time.

Y. H. Ma, *co-editor*
Worcester Polytechnic Institute
Worcester, MA

J. P. Ausikaitis, *co-editor*
Union Carbide Corporation
Waterbury, CT

CONTENTS

ADSORPTION AND DIFFUSION OF LIQUIDS IN SILICALITE USING HPLC

Y.H. Ma and Y.S. Lin ■ Department of Chemical Engineering, Worcester Polytechnic Institute, Worcester, MA 01609

Adsorption equilibrium constants and intracrystalline diffusivities of methanol, ethanol, acetone from aqueous solution and toluene, acetone in cyclohexane solution in silicalite crystals were determined using HPLC. The ease of using LC technique in the determination of adsorption and diffusion of liquids into very small zeolite crystal powders is demonstrated.

Adsorption from liquids into molecular sieve zeolites is finding increasing applications in separation and purification processes. The adsorption separations of liquid hydrocarbon mixtures (1, 2) and fructose/glucose mixture (3) by simulated moving-bed processes are two examples of these applications. However, despite its great importance either in practical industrial processes or in theoretical studies, diffusion as well as sorption equilibrium from liquids into zeolites, especially into zeolite crystals, has been scarcely studied due mainly to experimental difficulties. The well-stirred batch experimental techniques (static methods) have been often used for the measurements of liquid diffusion (4, 5, 6, 7, 8, 9) and sorption equilibrium (10, 11, 12) in porous materials. The essence of the static methods is to bring together known weight of solid and liquid into a vessel and to determine the transient concentration for diffusivity evaluation or allow equilibrium to be established and then determine the resulting change of the composition in the liquid for sorption equilibrium measurements. For the diffusion measurements, however, the static methods are generally inaccurate in determining the data, inconvenient in operating the experiments and time consuming during the experimental process.

Although chromatographic experimental techniques (dynamic techniques) have the advantages of simplicity, rapidity in obtaining data and easy adaptability to high temperature range, only for gases have these experimental techniques been extensively used for the measurements of diffusion and sorption in zeolites (13, 14, 15, 16, 17, 18). It was not until the 1970's that the chromatographic measurements were reported for liquid adsorption isotherms (e.g. 19, 20, 21, 22). Surprisingly, among the very limited work on the dynamic measurements of liquid diffusion in zeolites (23, 24), no work was carried out by means of commercially available liquid chromatography instruments.

Silicalite is a microporous crystalline silica with remarkable sieve properties (25). Although some limited studies have been reported on the properties of sorption and diffusion of gases (17, 26, 27, 28, 29) and sorption of liquids (12, 30), no study was reported on the properties of diffusion of liquids in silicalite crystal. There are increasing interests in applying the hydrophobic property of silicalite for organic removal from dilute aqueous solutions (e.g., separation of ethanol from dilute aqueous solution (31, 32, 33)). Thus, the studies of diffusion and sorption of organics in silicalite from aqueous solutions are important from both theoretical and practical points of view.

The growing availability of commercial high performance liquid chromatography (HPLC) instruments may also pave a way for the use of the chromatographic technique as a means for experimental studies of diffusion as well as sorption from liquids into zeolites. The main objective of the present study was to investigate the diffusion and adsorption of some organics in silicalite from aqueous solutions by the use of a commercially available HPLC instrument.

MATHEMATICAL MODEL

The intracrystalline diffusion, longitudinal dispersion and linear adsorption isotherm are considered in the modeling of an LC column packed with zeolite crystal powders. The differential mass balance over an element of the LC column and zeolite crystal results in two partial differential equations, which describe the mass transfer of the sorbate in the mobile phase of the column and the zeolite crystal pores. The mathematical model is based on the following assumptions:
(1) Uniform concentration profile in the mobile flow in radial direction due to small ratio of the diameter to the length of the column;
(2) Constant interstitial flow velocity along the length of the column due to the negligible pressure dependence of the liquid density;
(3) Fickian diffusion in the zeolite crystal with concentration and pressure independent intracrystalline diffusivity in the dilute solution;
(4) Spherical zeolite crystal particles with a uniform particle size.

The transport and adsorption phenomena involved in the process are described by the following partial differential equations:

Mass transfer in the mobile phase:

$$\frac{\partial C}{\partial t} + U\frac{\partial C}{\partial Z} + \frac{1-\varepsilon_b}{\varepsilon_b}\mathring{q} = D_L\frac{\partial^2 C}{\partial Z^2} \tag{1}$$

with the adsorption rate $\mathring{q}$ expressed by:

$$\mathring{q} = D_p\frac{3}{R}\left.\frac{\partial C_p}{\partial r}\right|_{r=R} \tag{2}$$

Diffusion in the zeolite crystal:

$$D_p\left(\frac{\partial^2 C_p}{\partial r^2} + \frac{2}{r}\frac{\partial C_p}{\partial r}\right) = \varepsilon_p\frac{\partial C_p}{\partial t} \tag{3}$$

It should be pointed out that the inclusion of ε_p in Eq. (3) is the result of the definition of the solute concentration which is based on the pore volume of the zeolite and appears to be more appropriate for a liquid system. All the symbols are identified in the nomenclature. For a column initially flowed with a pure solvent, the initial conditions are:

for the mobile phase

$$C\,(Z,t) = 0 \quad \text{at} \quad t = 0 \tag{4}$$

and for the zeolite crystal:

$$C_p\,(Z,r,t) = 0 \text{ at } t = 0 \tag{5}$$

The close-form boundary conditions (34) are taken for Eq. (1) as:

at the outlet of the column:

$$\frac{\partial C}{\partial Z} = 0 \quad \text{at} \quad Z = L \tag{6}$$

and at the inlet of the column:

$$C = \Gamma + \frac{D_L}{U}\frac{\partial C}{\partial Z} \quad \text{at } Z = 0 \tag{7}$$

where Γ is the square input pulse:

$$\Gamma = \begin{cases} C_o & (0 \le t \le t_{in}) \\ 0 & (\text{otherwise}) \end{cases} \tag{8}$$

The spherical symmetry condition is applied to the zeolite crystal:

$$\frac{\partial C_p}{\partial r} = 0 \text{ at } r = 0 \tag{9}$$

At the surface of crystal, we have

$$\frac{\partial C_p}{\partial r} = \frac{K_c}{D_p}\,(C - C_e) \text{ at } r = R \tag{10}$$

with the linear adsorption equilibrium relation:

$$C_p = K_e\,C_e \quad \text{at } r = R \tag{11}$$

Note the adsorption quantity derived from Cp in the above isotherm equation should be the absolute amount adsorbed in zeolite crystal (10, 12). The equilibrium constant in Eq. (11) can be calculated by the following equation derived from macroscopic mass balance over the entire LC column (35):

$$K_e = \frac{1}{v_p M} (Q\mu - V_m - \frac{V_1}{2}) \quad (12)$$

where the mobile phase volume V_m is defined as the bed void of the LC column.

Eq. (1) and Eq. (3) with the initial and boundary conditions as given above were solved by the orthogonal collocation numerical technique (36). The details of the numerical solution are given in (35). The direct time domain fitting method was used for the parameter evaluation.

EXPERIMENTAL

Apparatus and Materials

The experiments were carried out in a BECKMAN HPLC system, which consists of one model 421 system controller, two model 110 solvent metering pumps, one solvent mixer and one model 210 sample injector with 20 μl sample loop. A Hitachi model 100-40 UV-VIS spectrophotometer and an HP model 3390 reporting integrator were respectively used as the detector and recorder. Figure 1 shows the schematic of the HPLC system used in the experiments.

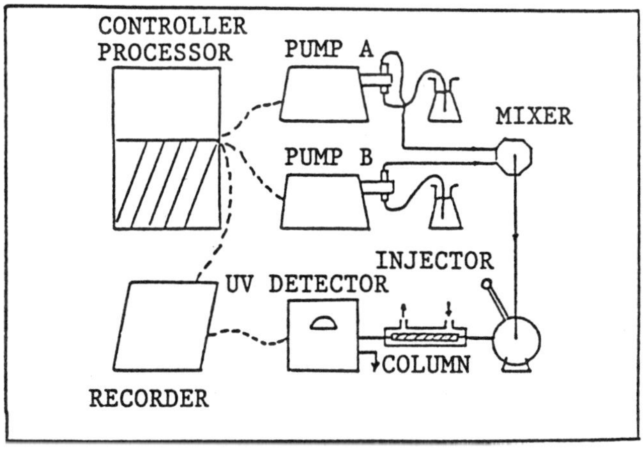

Figure 1. Schematic of HPLC apparatus.

The column consisted of a stainless-steel tube with polished inside wall, a pair of the column terminators with a minimum dead volume and two filters of 0.5 micron pore opening at the ends of the column. The column temperature was controlled by a water jacket in which the circulating water was supplied from a thermal bath.

The silicalite ("Linde" S-115) in spherical crystal powder form was obtained from Union Carbide Corp. The results of the particle size distribution measured by an Electrozone Celloscope particle size analyzer showed a very narrow crystal size distribution (in the range from 1.5 to 3.5 micron in diameter) with a mean diameter of 2.35 micron. The approximately spherical forms and the size of the crystals were verified by the Scanning Electron Micrographs. The physical properties and the chemical analysis of the silicalite are given in Table 1. All solvents used were of HPLC grade. In handling water, extreme care was taken to prevent water from being chemically or biologically contaminated in order to avoid poisoning silicalite by the carrier water.

Table 1 Specification of Silicalite Sample

Crystal Particle Radius R (micron)	1.17
Crystal Density ρ = (g/ml):	1.76
Pore Volume v_p (ml/g):	0.19
Crystal Porosity ε_p	0.344
Chemical Analysis (Wt%):	
SiO_2	98.6
Al_2O_3	0.44
Na_2O	0.01
C	0.06
NH_4	0.05

The zeolite samples were first activated at 300°C for about 15 hours and then cooled down in a dessicator to room temperature before being packed into the column. The stirred slurry packing methods (37) were tried in packing the column, but resulted in extremely high pressure drop (over 350 atm at Q = 0.3 ml/min for the column of 2 x 0.3) mainly because of the very small particle size and the low bed porosity. Consequently, the columns were packed by means of a dry-packing method, in which the zeolite powder in serial increments (typically each increment containing about 5 mg sample) was added into the column which was gently tapped on a hard surface. During the column packing, care was taken to ensure the uniform packing of the zeolite powder in the column. After the mass of the packed zeolite was measured gravimetrically, the packed column was then quickly connected to the HPLC system and filled with the carrier solvent.

After the column was packed with the zeolite powder, it was necessary to stabilize the column by flowing the carrier solvent through the column over an extended period of time. It was found that the column could be stabilized more effectively by varying the carrier flow rate for about six hours followed by a two day constant carrier flow rate. It is believed that the initial varying of the carrier flow rate facilitated the removal of residual gases in the particles while an extended period of constant carrier flow ensured the complete wetting of the particles. Furthermore, after an extended period of constant carrier flow, both the first moment, μ, and the HETP appear to reach the steady values as shown in Figure 2 which gives the first moment and HETP for methanol and ethanol, measured at the flow rate of Q = 0.5 ml/min as a function of total carrier water volume passed through the column. The increase in the second moment and decrease in the first moment may be the result of the continuous removal of the residual gas and wetting of the zeolite powder. It appears that after about 1 liter of carrier flow had passed through the column, both the first moments and HETP's reached the steady state values. All the experimental measurements for diffusion and sorption equilibrium therefore were carried out after the column was fully stabilized.

The preliminary experimental results also showed that too high a flow rate could result in unsatisfactory flow pattern such as channeling. For the column with the size of 2 x 0.2 cm, flow rates less than 3.0 ml/min gave a good flow pattern and proper pressure drop. In this flow rate range, a reasonable response peak could be obtained for a pulse injection; while in the higher flow rate range, however, sometimes multiple response peaks could be observed for pulse injection, due, probably, to possible channeling.

For the measurements of sorption equilibrium constants and intracrystalline diffusivities at dilute concentration range, the injection sample loop was first filled with a sample solution of a given sorbate concentration by a syringe. The sample was then injected into the column after a stable base line was obtained.

The height and corresponding time of the response peaks were directly read from the recording chart. For each peak at least fifteen points were read. The data were then input to a DEC-20 computer for the analysis. The response peak points were first interpolated by the cubic spline using IMSL routines ICSCCU and ICSEVU to give a continuous experimental response peak, from which the first absolute moment (mean) and second central moment (variance) were numerically calculated using the folowing equations:

$$\mu = \frac{\int_o^\infty Ct\ dt}{\int_o^\infty C\ dt} \tag{13}$$

$$\sigma^2 = \frac{\int_0^\infty C\ (t - \mu)^2\ dt}{\int_0^\infty C\ dt} \tag{14}$$

It should be noted that the peak fitted by the cubic spline essentially coincides with the experimentally measured peak.

The dead volume of the HPLC system was determined by plotting μ vs 1/Q for acetone with water as the carrier flow in an empty column and using the following mass conservation equation:

$$\mu = (V_c + V_d)\frac{1}{Q} \tag{15}$$

Substracting V_c from the slope of the μ vs 1/Q line gives the dead volume V_d (=0.022 ml), which includes the volume of the connecting tubes at the inlet and outlet of the column and the volume of the fittings. Corrections for the dead volume were made in the moment calculation. The bed porosity was calculated by the following equation:

$$\varepsilon_b = 1 - \frac{M}{\rho V_c} \tag{16}$$

Results and Discussions

The LC columns in different sizes were packed with the silicalite crystal powder in an attempt to obtain reliable LC column with good column performance. These columns were then used for the sorption and diffusion measurements. Table 2 gives the column parameters for the three typical silicalite columns used in the experimental measurements. The fact that the bed porosities and adsorption equilibrium constants among the three columns of different sizes are essentially the same indicates the good reproducibility in the column preparation.

As a further test of the reproducibility of the procedure for column packing, the pressure gradient is plotted against the interstitial velocity of the flow as shown in Figure 3. As expected, the pressure gradient data for all three columns fall on the same straight line according to the following correlation for low particle Reynolds number flow (38):

Table 2 Reproducibility of the column preparation

Column #	Size LxID (cm)	Silicalite Packed (mg)	Bed Porosity	K_e* (mL/mL)
1	2.01x0.21	64.0	0.469	7.31
2	2.01x0.20	62.5	0.464	7.36
3	2.30x0.40	267.0	0.464	7.21

*----for methanol+water+S-115 determined from means at 20°C

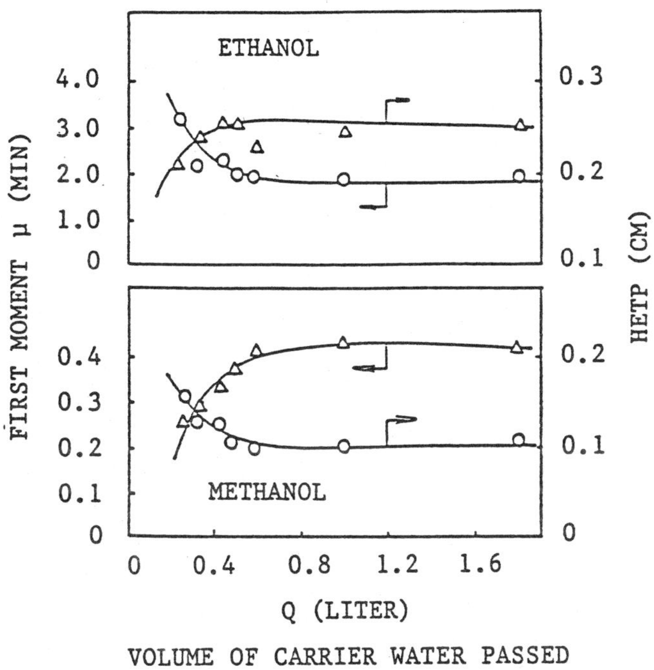

Figure 2. Stabilization of silicalite column.

$$\frac{\Delta P}{L} = f\frac{U}{(2R)^2} \quad (17)$$

The means and variances of the response peaks for methanol and ethanol for the column packed with the silicalite powder were determined at different flow rates. Th HETP was calculated by

$$\mathrm{HETP} = L\,\frac{\sigma^2}{\mu^2} \quad (18)$$

The means and HETP thus obtained are plotted as a functin of interstitial velocity in Figure 4. The two straight lines passing through the origin shown in Figure 4 (a) are consistent with the macroscopic mass balance Equation (12). The HETP vs interstitial velocity curves shown in Figure 4 (b) are typical for a well packed LC column (37). These evidences further show the high reproducibility of such mini LC columns packed with small zeolite crystal powders.

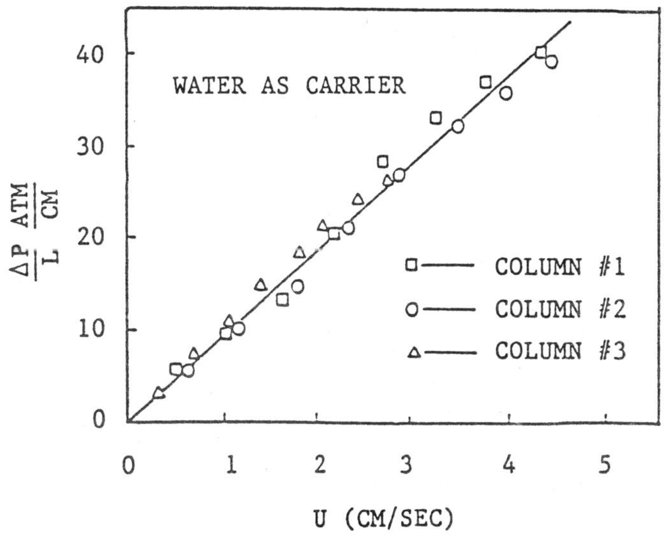

Figure 3. Pressure drop on silicalite columns.

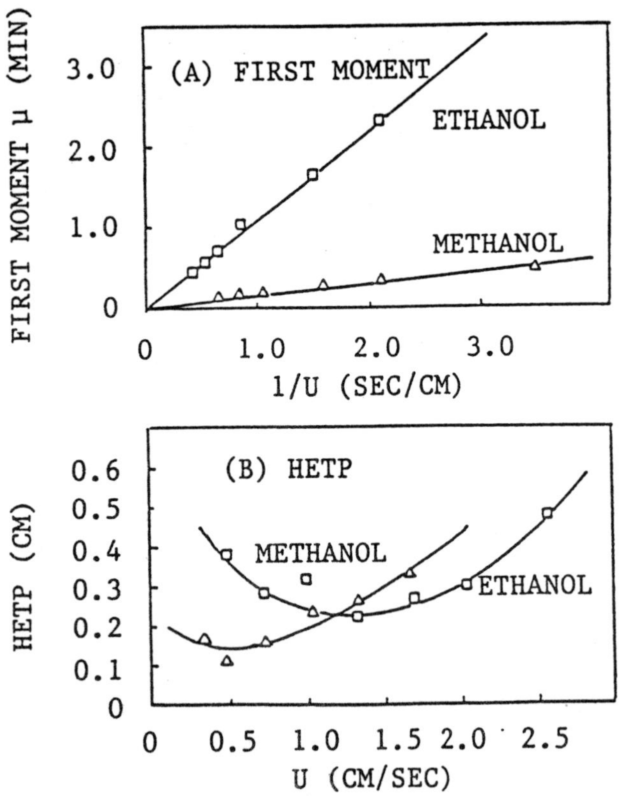

Figure 4. First moment and HETP of methanol and ethanol on silicalite column at 30°C.

The adsorption equilibrium constants and intracrystalline difusivities were determined from the response peaks of three different flow rates for a given temperature. The parameters used in the evaluation are given in Table 1 and Table 2. The axial dispersion coefficients were calculated using the following equation (39).

$$Pe = \frac{0.20}{\varepsilon_b} + \frac{0.011}{\varepsilon_b} R_e^{\ 0.48} \tag{19}$$

Owing to the very low particle Reynold number (Re is about 0.01), the external film mass transfer coefficients were determined by the following equation (40).

$$Sh = \frac{1.09}{\varepsilon_b} Re^{\ 0.33} Sc^{\ 0.33} \tag{20}$$

It should be pointed out that the sensitivity analysis performed on the current system showed that variations on these parameters have very small effects on the determination of the sorption equilibrium constants and diffusivities (35).

The sorption equilibrium constants were calculated using Eq. (12) from the means of the experimental response peaks at three different flow rates. The intracrystalline diffusivity was then calculated by one parameter fitting of an experimental response peak. The IMSL routine ZXSSQ using a finite difference Levenberg-Marguardt algorithm was used for the optimum parameter search.

The equilibrium adsorption and diffusion for methanol, ethanol and acetone in water (as the carrier) solution and toluene and acetone in cyclohexane (as the carrier) solution on the silicalite were studied. The experimental results of the adsorption equilibrium constants and diffusivities for the five liquid adsorption systems using the lowest injection sorbate concentration are given in Table 3.

The linearity of the equilibrium adsorption isotherms for the five adsorption systems was examined by injecting samples with different sorbate concentrations. Figure 5 shows the concentration dependency of the equilibrium constants at room temperature for the five adsorption systems. It is evident from Figure 5 (c), (d) and (e) that a linear adsorption isotherm for methanol + water, acetone + cyclohexane and toluene + cyclohexane can be assumed in the concentration range studied. On the other hand, for the systems of ethanol + water and acetone + water, the decrease in the equilibrium adsorption constants as the injection sorbate concentration is increased is evident from Figure 5 (a) and (b) indicating a non-linear behavior of the adsorption isotherms.

The typical experimental response peaks and the corresponding values computed from the model for methanol + water + S-115 system are shown in Figure 6 which shows an excellent agreement between the experimental and theoretical values. Similar good agreements are observed for the other two linear adsorption systems, acetone+cyclohexane and toluene+ cyclohexane. On the other hand, the extremely high value of K_e for acetone in water makes the assumption of a linear adsorption isotherm questionable even at very low injection sorbate concentration as evidenced by the somewhat poor agreement between the experimental and theoretical curves shown in Figure 7. A model taking into consideration of a non-linear adsorption isotherm is currently under development.

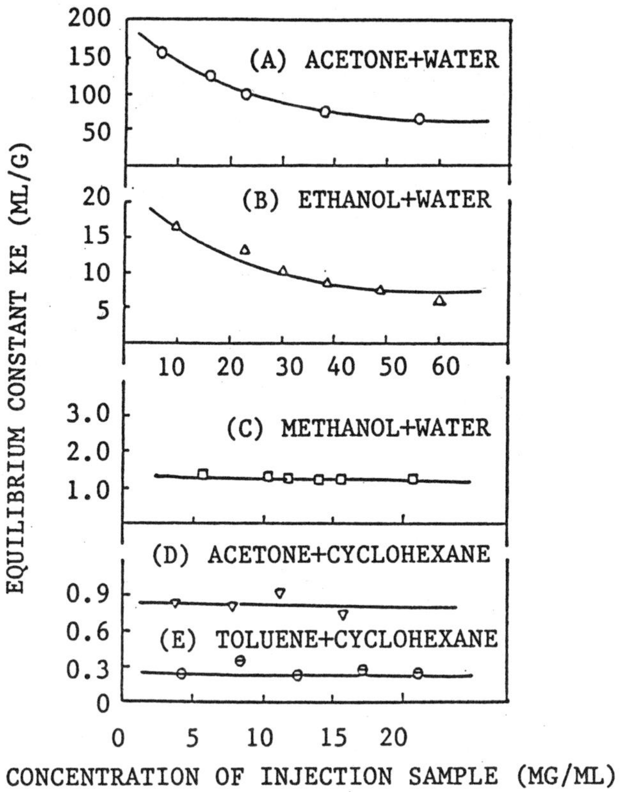

Figure 5. Concentration dependence of equilibrium constant.

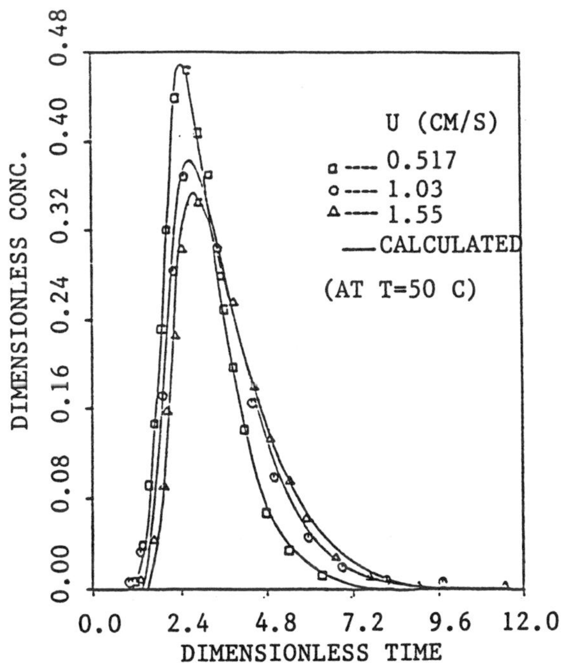

Figure 6. Response peaks of methanol+water+silicalite.

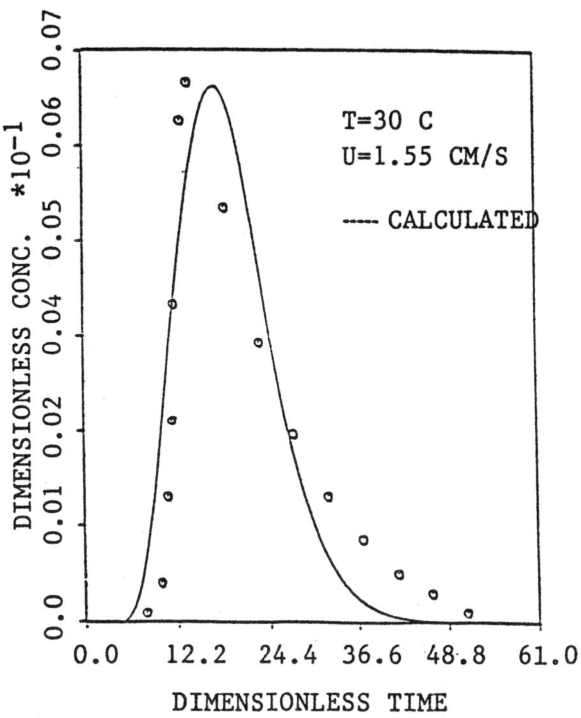

Figure 7. Response peak of acetone+water+silicalite.

From the equilibrium adsorption constants shown in Table 3, it can be seen that the adsorption capacity of methanol, ethanol and acetone in dilute aqueous solution increases in the following order:

methanol $<$ ethanol $<$ acetone

Due to its hydrophobicity, silicalite adsorbs alcohols as well as acetone much more preferentially than water (25, 32). The equilibrium adsorption constants for toluene and acetone in cyclohexane solution on silicalite are much smaller than those for the aqueous solution system (acetone+water). This is mainly due to the fact that cyclohexane has stronger affinity with silicalite than water (25, 26). The equilibrium adsorption constants of toluene and cyclohexane are about unity, indicating a similar affinity of the sorbate and solvent with silicalite. The fact that the equilibrium adsorption constant of acetone in cyclohexane solution is slightly larger than that of toluene is primarily due to the stronger affinity of acetone with the silicalite, which is consistent with the results from the aqueous solution systems.

The equilibrium adsorption constants for methanol and ethanol in aqueous solution on silicalite estimated from static experimental results reported by Milestone and Bibby (32) are also given in Table 3 for comparison. The good agreement between the results

Table 3 Adsorption Equilibrium Constants and Intracrystalline Diffusivities on Silicalite

Sorbate	Solvent	T (°C)	K_e(ml/ml)	$D_p x10^{10}(cm^2/s)$
		10.0	7.41	2.90
Methanol	Water	30.0	6.02 (7.0)[a]	3.40
		50.0	5.33	3.91
Ethanol[b]	Water	30.0	73.1 (71.0)[a]	2.24
Acetone[c]	Water	30.0	534.8	0.204
Acetone	Cyclo-	30.0	4.56	0.106
	hexane	50.0	3.81	0.455
Toluene	Cyclo-	30.0	1.22	0.797
	hexane	50.0	1.01	1.24

a----Estimated from Static Measurements at T = 20°C (32)
b----Lowest injection concentration=0.006 g/ml
c----Lowest injection concentration=0.0078 g/ml

measured by the static method and those obtained by the present LC method further demonstrates the validity of the LC technique for the determination of adsorption equilibrium and diffusion of liquids in zeolites.

The zeolite intracrystalline diffusion of liquids in the LC system investigated may be considered similar to the counterdiffusion of solute into zeolite crystal already saturated with a less adsorptive solvent (5, 6). The intracrystalline diffusivities thus obtained depend on the properties of the sorbate, the adsorbent and the carrier solvent. The somewhat lower diffusivity of ethanol than that of methanol is probably due to the difference in molecular sizes. The generally smaller diffusivities in cyclohexane than those in water appear to be due to the larger size of the solvent molecule cyclohexane which inhibits counterdiffusion of sorbate molecules.

The intracrystalline diffusivities shown in Table 3 are in the range of 10^{-10} to 10^{-11} cm^2/s. This is in the same order of magnitude as the counterdiffusion of cumene in NaY zeolite crystal with cyclohexane as solvent (5, 6). The diffusivities for the liquids reported here are also found to be in the same order of magnitude as those of vapor diffusion obtained by sorption uptake measurements (26). Bulow et al (41) recently reported that the liquid self-diffusion coefficients of n-decane in 5A zeolite crystals are about one order of magnitude greater than vapor phase diffusion coefficients. As a tentativae explanation, they attributed this difference to the preferential existence of liquid n-decane in the trans-conformation which diffuses faster than vapor n-decane which is preferentially in the gauch form. Althugh such an explanantion is plausible, a considerable amount of additional work would be needed in order to obtain a definitive answer to this appeared to be a rather unusual phenomenon. Haq (24) also found that the liquid diffusivities of some hydrocarbons in 10X zeolite crystals are in the same order of magnitude as the vapor sorption uptake diffusivities (42).

CONCLUSIONS

(1) An experimantal technique using a commercial HPLC instrument has been developed to determine the intracrystalline diffusion as well as sorption equilibrium of liquids in zeolite crystals. The determinations of adsorption and diffusion for the five adsorption systems of methanol, ethanol and acetone from aqueous solution and acetone and toluene from cyclohexane solution on silicalite crystals by the LC technique show the following advantages for the developed experimental technique:

(a) accuracy, rapidity and simplicity in obtaining data;
(b) easy adaptability to different temperature and pressure;
(c) avoidance of setting up a special experimental apparatus;
(d) use of small quantities of the adsorbent and adsorbate materials.

(2) The intracrystalline diffusivities were evaluated by direct time domain fitting of the LC response peaks. The mathematical model agrees well with the experimentally measured response peaks for the adsorption systems with linear adsorption isotherms. The intracrystalline diffusivities for the five adsorbates in water and cyclohexane solution were found to be in the same order of magnitude as the vapor intracrystalline diffusivities measured by the uptake method.

(3) The adsorption equilibrium constants of methanol and ethanol in aqueous solution in the silicalite determined by the LC technique agree well with those determined by the static method reported by the previous investigators.

NOMENCLATURE

C --- Concentration of the solute in the mobile phase [g/ml];
C_e --- Equilibrium concentration of the solute in the mobile phase [g/ml];
C_o --- Concentration of the solute in the injection sample [g/ml];
C_p --- Concentration of the solute in zeolite crystal pore [g/ml];
D_p --- Intracrystalline diffusivity based on the total area of zeolite crystal particle perpendicular to the diffusion direction [cm^2/s];
D_L --- Axial dispersion coefficient [cm^2/s];
f --- Coefficient defined in Eq. (17)[atm. s];
K_c --- External film mass transfer coefficient [cm/s];
K_e --- Adsorption equilibrium constant [ml/ml];
L --- Length of the packing in the column [cm];
M --- Mass of the zeolite packed [g];
Δp --- Pressure drop over the column [atm];
Q --- Volumetrical flow rate [ml/s] or [ml/min];
$\mathring{q}$ --- Adsorption rate on zeolite particle surface [g/ml.s];
R --- Radius of zeolite crystal particle [cm];
r --- Radius in zeolite crystal [cm];
t --- Time [s];
t_{in} -- Injection time, $t_{in} = V_l/Q$ [s];
U --- Interstitial flow velocity [cm/s]
V_m --- Volume of the mobile phase in the column [ml];
V_l --- Volume of the injection sample loop [ml];
V_c --- Volume of the empty column [ml];
V_d --- Dead volume of the LC system [ml];
V_p --- Pore volume of zeolite crystal [ml/g];
Z --- Distance along the column axial direction [cm];
Pe --- Peclet number $Pe = U\ (2R)/D_L$
Re --- Reynolds number, $Re = \rho_B \varepsilon_b U\ (2R)/\mu_B$
Sc --- Schmidt number, $Sc = \mu_B/\rho_B D^\circ_{AB}$
Sh --- Sherwood number, $Sh = K_c\ (2R)/D^\circ_{AB}$
ε_b --- Porosity of the packed column [ml/ml];
ε_p --- Porosity of zeolite crystal [ml/ml];
ρ --- Zeolite crystal density [g/ml];
μ --- First absolute moment [s];]
σ^2 --- Second central moment or variance [s^2]
Γ --- Square input pulse function.

LITERATURE CITED

1. Broughton, D.B. Neuzil, R.W., Pharis, J.M. and Brearly, C.S., Chem. Eng. Prog., 56 (4), 70 (1970).

2. Broughton, D.B. and Gembichi, S.A., "Fundamental of Adsorption", Ed. by Myers, A.L. and Belfort, G., Eng. Fund Conf. pp. 115-124 (1984).

3. Hashimoto, K., Adachi, S., Noujima, H. and Marjama, H., J. Chem. Eng. Japan., 16, 400 (1983).

4. Roberts, P.V. and York, R., Ind. Eng. Chem. Proc. Des. & Dev., 6, 516 (1967)

5. Satterfield, C.N. and Cheng. C.S., AIChE Symp. Series, 117, 67, 43 (1971).

6. Satterfield, C.N. and Cheng, C.S., AIChE J., 18, 724 (1972).

7. Satterfield, C.N. and Chiu, G.T., AIChE J., 20, 522 (1974).

8. Moore, R.M. and Katzer, J.R., AIChE J., 18, 816 (1972).

9. Choudhary, V..R. and Singh, A.P., Zeolites, 6, 206 (1986).

10. Kippling, J.J., "Adsorption from Solution of Non-Electrolytes", Academic Press, pp. 5-31 (1965).

11. Everett, D.H., "Fundamentals of Adsorption", Ed Myers, A.L. and Belfort, G., Eng. Fund. Conf., pp; 1-35 (1984).

12. Ma, Y.H. and Lin, Y.S., AIChE Symp. Series, 242, 81, 39 (1985).

13. Ma, Y.H. and Mancel, C.P., "Molecular Sieves", Adv. Chem. Ser., No121, 392(1973).

14. Ruthven, D.M. and Kumar, R., Canadian J. Chem. Eng., 57, 342 (1979).

15. Shu, L.-K. P. and Haynes, H.W., AIChE J., 27, 81 (1981).

16. Hyun, S.H. and Danner, R.P., AIChE Symp. Series, 219, 78 (1982).

17. Chiang, S.C., Dixon, A.G. and Ma, Y.H., Chem. Eng. Sci., 39, 1461 (1984).

18. Boniface, H.A. and Ruthven, D.M., Chem. Eng. Sci., 40, 1401, (1985).

19. Sharma, S.C. and Fort Jr., T., J. Colloid and Interface Sci., 43, 36 (1973).

20. Wang, H.L., Duda, J.L. and Radke, C.J., J. Colloid and Interface Sci., 66, 153, (1978).

21. Jacobson, J., Frenz, J., Horvath, C., J. Chromatog., 316, 53 (1984).

22. Jacobson, J., Frenz, J., Horvath, C., Ind. Eng. Chem. Res., 26, 43 (1987).

23. Lee, T.K. and Ruthven, D.M., Ind. Eng. Chem. Fundam., 16, 290 (1977).

24. Haq, N., "Chromatographic Study of Diffusion and Sorption for Gases and Liquids in Type A and 13X Zeolites", Ph.D. Thesis, Univ. of New Brunswick, Canada (1983).

25. Flanigen, E.M., Bennett, J.M., Rgrose, R.W., Cohen, J.R., Patton, R.L., Kirchner, R.M. and Smith, J.V., Nature (London), 257, 512 (1978).

26. Wu, P.D., Debeve, A. and Ma, Y.H., Zeolites, 3, 117 (1983).

27. Stack, H., Thamm, H., Jachen, J., Fielder, K. and Schirmer, W., Proc. 6th. Internat. Zeol. Conf., Ed. A. Bisio and D.H. Olson, 225 (1983).

28. Lechert, H. and Schweitzer, W., Proc. 6th Internat. Zeol. Conf., Ed. A. Bisio and D.H. Olson, 210 (1983).

29. Paravar, A. and Hayhurst, D.T., Proc. 6th Internat. Zeol. Conf., Ed. A. Bisio and D.H. Olson, 217 (1983).

30. Narita, E., Horiguchi, N. and Okabe, T., Chem. Lett., 787 (1985).

31. Oulman, C.S. and Chriswell, C.D., U.S. Pat., No 4277635 (July 7, 1981).

32. Milestone, N.B. and Bibby, D.M., J. Chem. Tech. Biotechnol., 31, 732 (1981).

33. Bui, S., Verykios, X. and Mutharasan, R., Ind. Eng. Chem. Process. Des. Dev., 24, 1209 (1985).

34. Butt, J.B., "Reaction Kinetics and Reactor Design", Prentice Hall, p. 291(1980).

35. Ma, Y.H. and Lin, Y.S., in preparation, (1987).

36. Villadsen, J.V. and Stewart, W.E., Chem. Eng. Sci., 22, 1483 (1967).

37. Snyder, C.R. and Kirkland, J.J., "Introduction to Modern Liquid Chromatography", 2nd Ed., Wiley p. 216, pp. 168-172 (1979).

38. Bird, R.B., Steward, W.F. and Lightfoot, E.N., "Transport Phenomena", Wiley pp. 196-201 (1960).

39. Wen, C.Y., Fan, L.T., "Models for Flow Systems and Chemical Reactors", Dekker, p. 165 (1975).

40. Wilson, E.J., Greankoplis, C.J., Ind. Eng. Chem. Fundam., 5, 9 (1966).

41. Bulow, M., Struve, P. and Rees, L.V.C., Zeolites, 5, 113 (1985).

42. Ruthven, D.M. and Doetsch, I.H., AIChE J., 22, 882 (1976).

SURFACE ENERGETICS IN THE ADSORPTION OF HYDROGEN CHLORIDE ONTO TRANSITION ALUMINAS

Hubert L. Fleming ■ Alcoa Laboratories, Aluminum Company of America, Alcoa Center, PA 15069

Adsorption of species onto porous solids is often treated macroscopically as a homogeneous process. Interaction between the adsorbate and a solid is usually characterized by a single parameter, typically an equilibrium coefficient derived from isotherm measurements (1,2). This derivation is implicit in the Langmuir model, as well as other widely employed constructs for the adsorption process. In regimes where chemisorption predominates, these models exhibit severe limitations, in that the adsorption process is quite heterogeneous. Because of nonhomogeneous surface energetics of the solid, it is quite difficult to predict adsorption as a function of adsorbate concentration, temperature, and pressure-system variables that extensively influence the affinity of the surface for the adsorbate as a function of the surface coverage.

Only recently have investigators begun to appreciate effects of surface energetics on adsorption phenomena. Jaroniec and others (3,4) have predicted adsorption behavior as a function of theoretical surface potential functions. There is an intrinsic question, however, as to what surface energetics of a solid in a particular system are, and how they can be characterized.

The interaction of gaseous hydrochloric acid with transition aluminas has been examined. Materials with relatively well characterized surfaces have been modified via sodium hydroxide adsorption in an attempt to understand HCl adsorption as a function of the strength and number of Lewis and Bronsted acidity of the surface. Using a technique employing simultaneous mass and calorimetric measurements of the adsorption process on the solid (described elsewhere) (5,6), surface energetics during adsorption of HCl on heterogeneous alumina surfaces are characterized, and related to bulk adsorption phenomena.

EXPERIMENTAL PROCEDURES

Three materials were used in experiments reported here. Alcoa S-100, a commercial transition alumina which is primarily amorphous

Aluminum Company of America, Pittsburgh, Pennsylvania

with some gamma and chi alumina components, was the base material. S-100 is Material 1 in the tabulation of physical and chemical properties of adsorbents presented in Table 1. One fourth inch diameter spheres of S-100 were soaked at 20°C in a stirred tank of aqueous NaOH solution and then oven dried in flowing air at 110°C overnight. Material 2, 0.58 wt.% surface sodium (reported as Na_2O), was obtained with a 1.0 wt.% NaOH solution, soaked for one hour. The 3.5 wt.% Na_2O loading was obtained with a 5.0 wt.% solution with a four hour contact time. All materials were soaked in excess NaOH (> 1 L/100 g). The spheres were then crushed and sized to -100 +800 Tyler mesh, characterized (by XPS, TGA, XRD, and FTIR), and used in the adsorption experiments.

Samples of 1- to 50 mg mass were inserted into a microreactor system (custom built by S.E.T.A.R.A.M., Lyon, France). Simultaneous Thermogravimetric Analysis-Differential Thermal Analysis (TGA-DTA) was obtained on samples in real time. The cell design is shown in Figure 1. Absolute mass sensitivity in the configuration used for these studies was 0.4μg (relative mass sensitivity, 4×10^{-8}). DTA was semiquantitative following the procedure of Wendlandt (7). K (486.41 kcal/uV-min at 25°C in N_2) was calibrated by the heat of fusion of various metals, including gallium for low temperature sensitivity. Measured DTA sensitivity was 5 mcal/g Al_2O_3 for HCl adsorption, corresponding to ±0.2 kcal/mole HCl adsorbed. TGA was zeroed using an offset reference chamber containing 0.2 m^2/g alpha alumina as a standard. System configuration is shown in Figure 2. Data collection (TGA, DTA, chamber and sample temperatures, system pressure, HCl partial pressure, HCl and N_2 flow rates) was made every 0.1 msec by two HP86A computers. Data manipulation packages were designed and written on a VAX 780, where the raw data was subsequently transferred. Gas mixing and flow introduction was by a network of Teledyne-Hastings mass microflowmeters (0 to 120 cc/min total flow at STP). Union Carbide Ultra High Purity gases (lecture bottles) were employed in this study. Online Balzers diffusion pumps (backed up by a Lebold-Hercous turbomolecular pump) allowed controlled chamber desorption and evacuation.

A typical run procedure is depicted in Figure 3. Alumina powder was pretreated in-situ at 110°C, one mtorr pressure until change in weight loss was not observed (typically one to two hours). After the temperature of the system was stabilized at 25°C, flowing N_2 (110 cc/min) was introduced, and the TGA/DTA was zeroed. HCl was introduced at a desired concentration by controlling the ratio of HCl to N_2, while maintaining a constant total flow. For the study reported here, HCl concentrations of 5×10^{-5} (five ppm), 10^{-5}, 5×10^{-4} 10^{-4}, 10^{-2}, and 10^{-1} bar were examined. Pure HCl data was obtained in a separate packed bed reactor, due to flow constraints on the microreactor. The desired HCl concentration was maintained under flow until steady state was reached on both the TGA (mass uptake) and DTA (heat of adsorption). Once steady state was reached, the concentration was stepped to the next desired higher value. Following steady state adsorption at 0.1 bar HCl, gas flow was discontinued and the system evacuated to x 10^{-3} torr. Following desorption of the weakly bound states, a linear thermal ramp (25 to 1000°C at 3°C/min) was performed to obtain Temperature Programmed Desorption (TPD) data (to be reported in a subsequent communication). Following TPD, the sample was exposed to atmospheric pressure and ambient temperature, and the run terminated.

RESULTS

Isotherms for each of the three materials examined at 25°C are given in Figure 4. Monolayer surface coverage for all materials is approximately 2.0 wt.% at very low

HCl particle pressures (<100 ppm), and is not dependent upon the partial pressure. Considerable difference in HCl uptake among materials is observed in the range of HCl partial pressure from 10^{-4} to 10^{-2} bar. Adsorption capacities increase for the low sodium materials linearly over this range, then approach an asymptotic limit at around 7.0 wt.% at HCl saturation.

Both materials treated with NaOH exhibit increase HCl capacities at the higher partial pressures of HCl, with a strong Type 1 modified isotherm observed for the alumina with 3.5 wt.% Na_2O. Saturation was not reached for this material. Equilibrium adsorption of HCl on the 0.58 wt.% Na_2O alumina was intermediate to the other two. A transition is seen for increased adsorption in the intermediate HCl partial pressure region for this material, yet it still maintains a similar coverage at $P/P_o=1$ (defined as pure HCl at 1.0 bar total system pressure).

Minimal desorption of weakly bound states was observed for any of the aluminas. Vacuum desorption of HCl is summarized in Table 2 for two levels of HCl partial pressure, taken by separate experiments. Less than five percent of the HCl adsorbed was isothermally desorbed, even for the alumina with 3.5% Na_2O which continued to exhibit a linear uptake at high HCl partial pressures. The implication of this observation is that physisorption is unlikely as a mechamism in HCl adsorption, and more specific interactions, notably hydrogen bonding and chemisorption, must account for the uptake behavior even at very high P_{HCl} at 25°C.

Differential enthalpies of HCl adsorption onto the aluminas are given in Figure 5 as a function of HCl uptake. Relatively high heats, on the order of -30 kcal/mole HCl, are seen for initial coverage on the native alumina surfaces. Virgin S-100 alumina exhibits slightly higher initial heats than impregnated aluminas by one to two kcal at -32.1 kcal/mole. Surfaces are seen to be quite neterogeneous in all cases, with enthalpies of 10 to 15 kcal/mole at coverages approaching saturation. S-100 samples impregnated with NaOH exhibit sustained coverage of sites with high (>25 kcal/mole) interaction energies over a much broader range of coverage. For none of the materials did the differential enthalpies drop down into the region of one to ten kcal/mole, indicative of minimal weakly bound physisorbed complexes.

Because total surface coverage at saturation varies for the three aluminas, comparison of the energetic heterogeneity is better done by normalization of the total uptake to a saturation coverage. The data of Figure 5 is replotted in Figure 6 as a function of surface coverage, θ. $\theta=1$ corresponds to saturation coverage (defined as $P/P_o=1.0$, or earlier for two of the materials). Two additional features are observed when the data is plotted in this manner. First, the isotherm of alumina impregnated at the intermediate level with NaOH, exhibits HCl adsorption features which are composites of the isotherms virgin S-100 and the 3.5% Na_2O alumina. High interaction energies, on the order of 28 to 30 kcal/mole, are maintained in the 0.58 wt.% alumina, even up to $\theta=0.4$, corresponding to 40% of surface coverage. This value is greater than would be predicted by a linear interpolation of high interaction energies as a function of Na_2O content. The expected value was similar to the value for S-100. Behavior of the 0.58 wt.% Na_2O alumina is similar to S-100 at high coverage, $\theta > 0.6$.

A second feature not obvious in the previous plot is the hyperbolic shape of the enthalpy curve toward a lower limit for the S-100 with no NaOH and low NaOH. In contrast, S-100 with 3.5 wt.% Na_2O continues to exhibit a parabolic drop-off in site energy at high surface coverage. This observation is thermodynamically consistent because alumina did not reach a limit in weight gain ($\theta<1$), even when contacted with pure HCl.

Previous data of simultaneous mass and calorimetry can also be manipulated to obtain distribution of surface site potentials, as characterized by differential enthalpies of adsorption, as a function of number of sites corresponding to these energies. Site-energy distribution is determined by differentiation of differential enthalpies of adsorption by number of sites (essentially a derivative curve of Figure 5), and is plotted in Figure 7. Data points are omitted for clarity. Units for the ordinate axis are in terms of moles ^{2}HCl/g Al_2O_3-kcal, which is the number density of sites at a given potential energy for desorption.

Two types of sites are determined for S-100. High energy sites are characterized with WH_{AD} = 25-32 kcal/mole, and are widely distributed. Low energy sites ($-\Delta H_{AD}$ = 12-15 kcal/mole) are also present in significant density with a peak value at 11.8 kcal/mole. The NaOH-impregnated aluminas exhibit slightly weaker strong sites (<30 kcal/mole), but possess high number densities, much more than S-100. The decrease in number density for sites less than 25 kcal/mole is continuous with a significant density of lower energy sites observed for the intermediate Na_2O material.

As seen in Figure 7, data plotted in this format has discontinuous curves at both the low and high values of the differential enthalpy of adsorption. This is not realistically correct since a continuum of sites exists, particularly at low energy where it is difficult to probe experimentally. Investigation of low energy sites would require adsorption at lower temperatures in order to make these sites thermodynamically favorable for adsorption. Subsequently, these can then be observed and characterized experimentally.

DISCUSSION

Effect of Pore Distribution

Quantification of surface energetics in the interaction of hydrochloric acid with a transition alumina impregnated with sodium hydroxide was the objective of this study. Ideally, the pore size distribution and shape would remain constant after impregnation. As shown in Figure 8, impregnation of S-100 with sodium hydroxide resulted in local dissolution of the micropore structure, and subsequent formation of larger average pores with values of 80 to 100Å. The total pore volume, measured by the combination of He and Hg porosimetry, decreased from 0.54 to 0.43 cc/g for S-100 and 3.5 wt.% Na_2O alumina through this chemical sintering process.

Increased macropore structure manifests itself in the isotherms (Figure 4) by the nonasymptotic behavior of HCl adsorption at high P_{HCl} for impregnated aluminas. There is a continual increase in HCl uptake at $P_{HCl} > 10^{-2}$ bar, which significantly exceeds 7.0 wt.% HCl. If the free van der Waals diameter of HCl is 4.5Å, and most of the micropores present are 20Å or less in diameter, then it is not possible for more than monolayer coverage of HCl to be present. This results in a finite, asymptotic capacity of HCl, as observed for S-100. For impregnated materials, having more pores greater than 20Å, significant multilayering due to hydrogen bonding and condensation within the pores can be envisioned. The result is greater HCl uptake at high P_{HCl}, as observed in isotherms for $P_{HCl} > 0.01$ bar.

A change in mechanism is consistent with the decrease in differential heats of adsorption for HCl (Figure 5) in the range of 6 to 8 wt.% HCl from 25 to 30 kcal/mole, down to 10 to 20 kcal/mole. Because the value of $-\Delta H_{AD}$ remains above 10 kcal/mole, it is likely multilayering results from hydrogen bonding, rather than simple pore condensation. Further evidence for strongly bound multilayers is the small weight loss from facuum desorption data (Table 2), where minimal HCl desorption was found to be isothermally reversible, even for the relatively macroporous alumina of high Na_2O content. Noninteractive pore filling would have resulted in much more desorption under reduced pressure.

Hydrogen chloride uptake in the multilayer region (Figure 4) is 10% greater for the intermediate sodium-impregnated alumina than for S-100. Further, uptake varys from 12 to 96% greater for the high-sodium material. Although S-100 possesses the greatest total pore volume, not all of the porosity is available for condensation. In the large meso and macropore regions, for example, as defined by pores 50Å (Figure 8), S-100 exhibits the lowest pore volume, with the high-sodium alumina having the largest pore volume. This is further manifested in the average pore size of the alumina, with the size increasing with sodium content. The high degree of multilayering at saturation for the 3.5 wt.% Na_2O alumina is puzzling, however.

Effect of Surface Energetics

For alumina S-100, the saturation capacity at monolayer coverage was found to be 2.2 wt.% HCl. This value is determined as the asymptotic coverage in Figure 4, and is the base amount which is not reversible without an imposed energy of desorption. This figure is consistent with the asymptotic function of uptake versus P_{HCl} in Figure 4, and with the differential enthalpy over the same function in Figure 5. The Langmuir equation calculates the cross sectional area of a surface HCl molecule 34.3 $Å^2$. Given that the cross sectional area of each coordinatively unsaturated aluminum ion (with adjacent oxygens) is approximately 16$Å^2$ (depending upon the particular transition form). This is consistent with the observation of one HCl molecules adsorbed for every two aluminum cations. Given the free van der Waals area of an HCl molecule is 16.6$Å^2$, it is unlikely any greater coordination of HCl to the surface could be predicted. As depicted in Figure 9, it does not appear possible to associate a chloride ion with every aluminum cation. This finding is consistent with that of Bailey and Wightman who measured an effective van der Waals area for HCl on γ-alumina of 33.4$Å^2$ (8).

Over the region of monolayer coverage on S-100, the enthalpy of adsorption varies only by two to three kcal/mole. The slight decrease is probably due to weak interaction of adjacent sites, a result of possible cation charge cloud decrease with continual delocalization with each additional chloride ion. In the intermediate region above monolayer coverage and prior to approach of the asymptotic adsorption limit, there is a sharp decrease to 12 to 13 kcal/mole for enthalpies of adsorption. Above monolayer coverage on S-100, two to three multilayers are formed primarily by hydrogen bonding.

Given relatively high heats of adsorption at low surface coverage, some speculation can be made regarding the mechanism of surface attachment of the HCl. It is generally accepted the chloride attaches to a native transition alumina surface by covalent bonding to a coordinatively unsaturated aluminum cation functioning as a Lewis acid. Both our own unpublished spectroscopic studies, as well as those in the literature are consistent on the mode of chemical attachment (9, 10). The question remains whether there is disassociation of the HCl, followed by attachment of the chloride ion to the aluminum, with the subsequent formation of an adjacent surface hydroxyl.

Assuming a gas phase HCl bond is broken (103 kcal/mole) and Al-Cl and O-H bonds form (-55.4 and -50.9 kcal/mole, respectively), the process is relatively thermally neutral at ΔH = -3.3 kcal/mole. Assuming no disassociation, the process has a value of ΔH = -47.6

kcal/mole, a much closer value to that observed experimentally. Data as a function of adsorption temperature needs to be obtained to determine the activation energy of the process in order to gain further insight.

Surface energetics were similar for native alumina and for sodium-impregnated materials in the region of submonolayer coverage. Saturation capacities as seen in Figure 4 are slightly less at under 2.0 wt.%, yet there is less surface area and micropore volume. The calculated van der Waal area of HCl was 36.4 and 38.2Å^2, for materials number two and three, respectively. The reduced normalized monolayer coverage is most likely due to a reduction in the number of Lewis acids for adsorption. Some poisoning of the coordinatively unsaturated aluminum cations is anticipated as a result of sodium ion blockage. In addition, ion exchange of sodium for protons of surface hydroxyl groups can occur. This most likely occurs with the subsequent formation of additional hydroxyls via the disassociative adsorption of the NaOH, since it was observed (Table 1) the total hydroxyl content increased from 5.0 wt.% to 8.3 wt.% through the impregnation process. Adsorption of HCl via these additional surface hydroxyls (Bronsted sites) was not considered a significant mechanism for adsorption, since no increased adsorption capacity was observed. Also, the total surface acidity (Bronsted and Lewis), as measured by ammonia adsorption to decrease at increasing sodium content (Table 1).

Sodium adsorption/poisoning of strong Lewis acid centers resulted in a decrease in ΔH at θ=0 of two to four kcal/mole, best seen in Figure 6. The total number and strength of adsorption sites were slightly decreased. The resolution of the experiment did not show a difference between the 0.58 wt.% Na_2O and the 3.5 wt.% Na_2O aluminas up to monolayer coverage.

In the regime of two to three multilayers ($P_{HCl} > 10^{-4}$ bar), both the number and strength of sites for adsorption in the sodium-impregnated aluminas increased. Heats of adsorption over 20 kcal/mole were maintained through one additional layer of HCl. Hydrochloric acid was most likely being coordinatively attached to the surface sodium species. Although the site potential was lower than that of an aluminum cation (lower charge/mass ratio), at high HCl partial pressures a second chemisorption mechanism utilizing the surface sodium appears to become significant. Given the total quantity of this second strongly bond layer almost exactly corresponds to monolayer coverage (for the 3.5 wt.% Na_2O, and 80% for the 0.58 wt.% Na_2O), it is possible HCl attachment was through second level sodium projected out from the surface via ion exchanged hydroxyl groups. The stearic limitation of attachment to lattice aluminums would then be removed. Site potential for adsorption decrease from 30 to under 20 kcal/mole.

Bulk acid/base reaction of HCl with molecular sodium hydroxide is not considered a possible mechanism for surface interaction. Not only is bulk NaOH not observed spectroscopically or by X-Ray Diffraction measurements, measured heats of adsorption are far too low and inconsistent as a function of sodium content. Mass balances of chloride reacted with the various levels of sodium are also grossly inconsistent, as are the activation energies for reversible reaction (not presented here).

At high P_{HCl}, saturation of HCl on sodium-modified aluminas may not be complete. Enthalpies of adsorption remain relatively high, 14 to 16 kcal/mole, indicative of a hydrogen bonding process. However, the values have not yet approached that of native S-100, indicative of complete pore filling. Given that the latent heat of vaporization is approximately five kcal/mole, the data suggest that the multilayering is more complex than simple condensation with observed heats of adsorption in this region varying from 10 to 15 kcal/mole, more association, such as hydrogen bonding must be occurring.

SUMMARY

The interaction of hydrochloric acid with transition alumina surfaces modified by sodium hydroxide was considered. Adsorption proceeds by an exothermic process with monolayer coverage at 2.2 wt.% HCl on the native alumina. Sodium-modified aluminas exhibit lower affinities for HCl adsorption, both in terms of number and strength of sites.

On the alumina, adsorption occurs via the route of monolayer chemisorption of HCl, possibly disassociated, followed by pore condensation with some stronger interaction, presumably hydrogen bonding. Surface energetics indicate quite distinct sites, with a bimodal site-energy distribution corresponding to these two regimes of interaction. Minimal reversibility is observed in the adsorption process.

For sodium-impregnated aluminas, a third mode of HCl adsorption is proposed, where HCl is bound to surface sodium species. The attachment is strong and specific, with binding energies on the order of 20 to 25 kcal/mole. The result is increased adsorption capacity of HCl at intermediate partial pressures of HCl when compared with alumina. Additional hydrogen bonded adsorption of HCl is exhibited by sodium-impregnated aluminas because of an increase in macropore volume in the 60 to 100Å region.

Surface energetics from each regime of interaction are quite distinct, and are characterized by plotting site-energy distributions as a function of coverage. Information concerning the mechanisms of the adsorption processes require additional data, particularly surface spectroscopy and adsorption measurements at various temperatures.

LITERATURE CITED

1. J. M. Smith, Chemical Engineering Kinetics, 3rd Edition, McGraw-Hill, New York (1981).
2. J. J. Carberry, Chemical and Catalytic Reaction Engineering, McGraw-Hill, New York (1976).
3. A. W. Marczewski et al., J. Colloid Interface Sci., 109 (2), 310 (1986).
4. P. Brauer and M. Jaroniec, J. Colloid Interface Sci., 108 (1), 50 (1985).
5. H. L. Fleming, Proc. North Amer. Thermal Anal. Soc. Mtg., Philadelphia (September, 1984).
6. H. L. Fleming, Pittsburgh Conf. Appl. Spec. Anal. Chem., Atlantic City (March, 1985).
7. W. W. Wendlandt, Thermal Methods of Analysis, Interscience Publishers, New York (1964).
8. R. R. Bailey and J. P. Wightman, J. Coll. Interface Sci., 70 (1), 112 (1979).
9. M. Tanaka and S. Ogasawara, J. Catalysis, 16, 157 (1970).
10. V. F. Vozdvizhenskii et al., Izvestiya AN Kaz CCR. Seria Khimicheskaya, (USSR), 3 (1976).

Table 1. Physical and chemical properties of aluminas.

MATERIAL	1	2	3
Surface Na Content* (As wt% Na_2O)	0	0.58	3.5
Total Impurities (wt%)	0.42	0.42	0.42
Phase (XRD)	Amorphous, weak γ/χ	Amorphous, weak γ/χ	Amorphous, weak γ/χ
Surface Area (B.E.T. N_2)	327	317	302
Total Surface Acidity (wt% NH_3)**	0.47	0.38	0.35
Hydroxyl Content (TGA, 110°C-1000°C)	5.0	6.5	8.3
Total Pore Volume (Hg, He, cc/g)	0.54	0.49	0.43
Average Pore Diameter (Å, at 50% Volume)	48	49	51
He Density (g/cc)	3.13	3.11	3.08
Particle Size (Tyler Mesh)	-100 +180	-100 +180	-100 +180

*Excludes 0.35 wt% interstitial Na_2O

**100 ppm NH_3 in flowing Ar, 20°C, 1 atm

Table 2. Isothermal vacuum desorption.

	AT 100,000 PPM HCl			AT 100% HCl		
Material	Capacity (wt%)	Amount Desorbed (wt%)	% of Total Desorbed	Capacity (wt%)	Amount Desorbed (wt%)	% of Total Desorbed
S-100	6.78	0.22	3.2	6.80	0.24	3.5
S-100 W/0.58% Na_2O	7.41	0.23	3.1	7.59	0.24	3.2
S-100 W/3.5% Na_2O	8.78	0.31	3.5	11.29	0.48	4.3

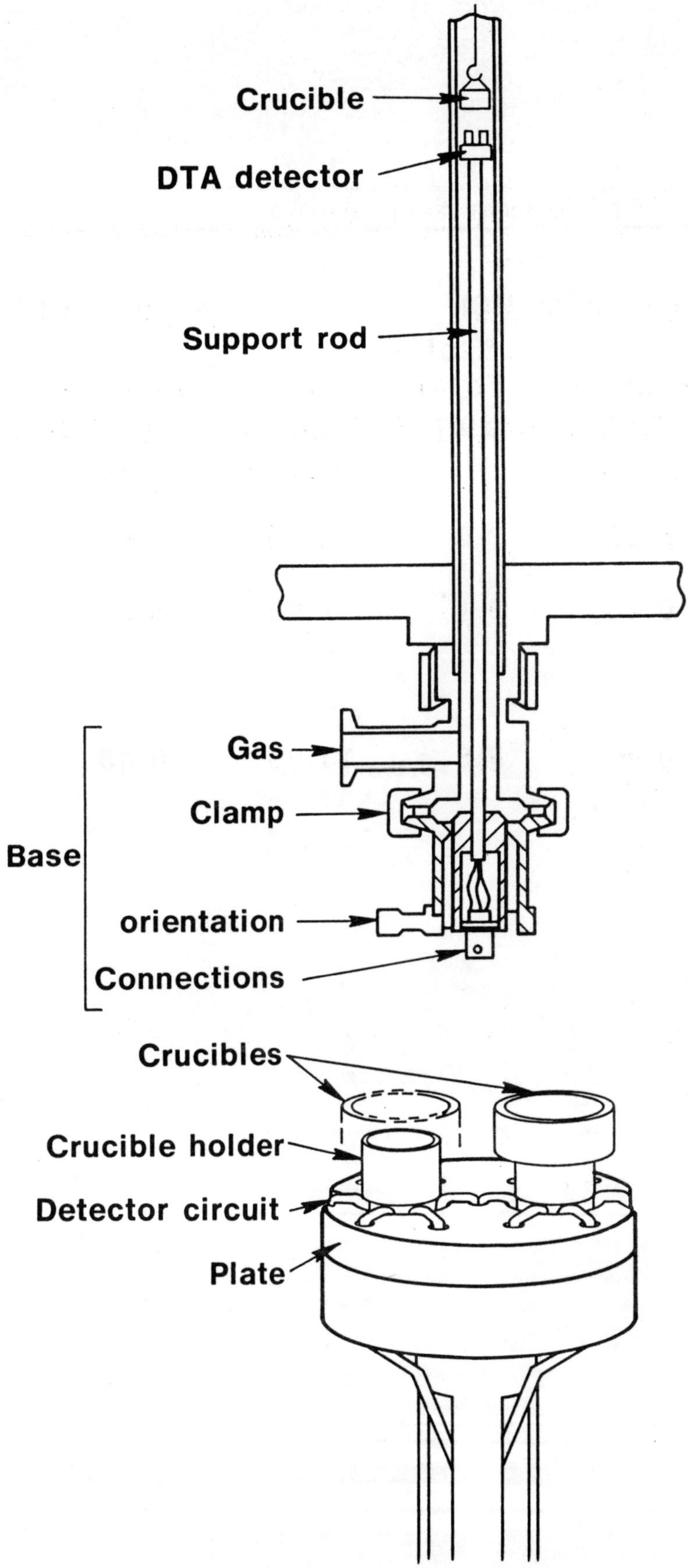

Figure 1. Simultaneous TGA-DTA cell (from S.E.T.A.R.A.M.).

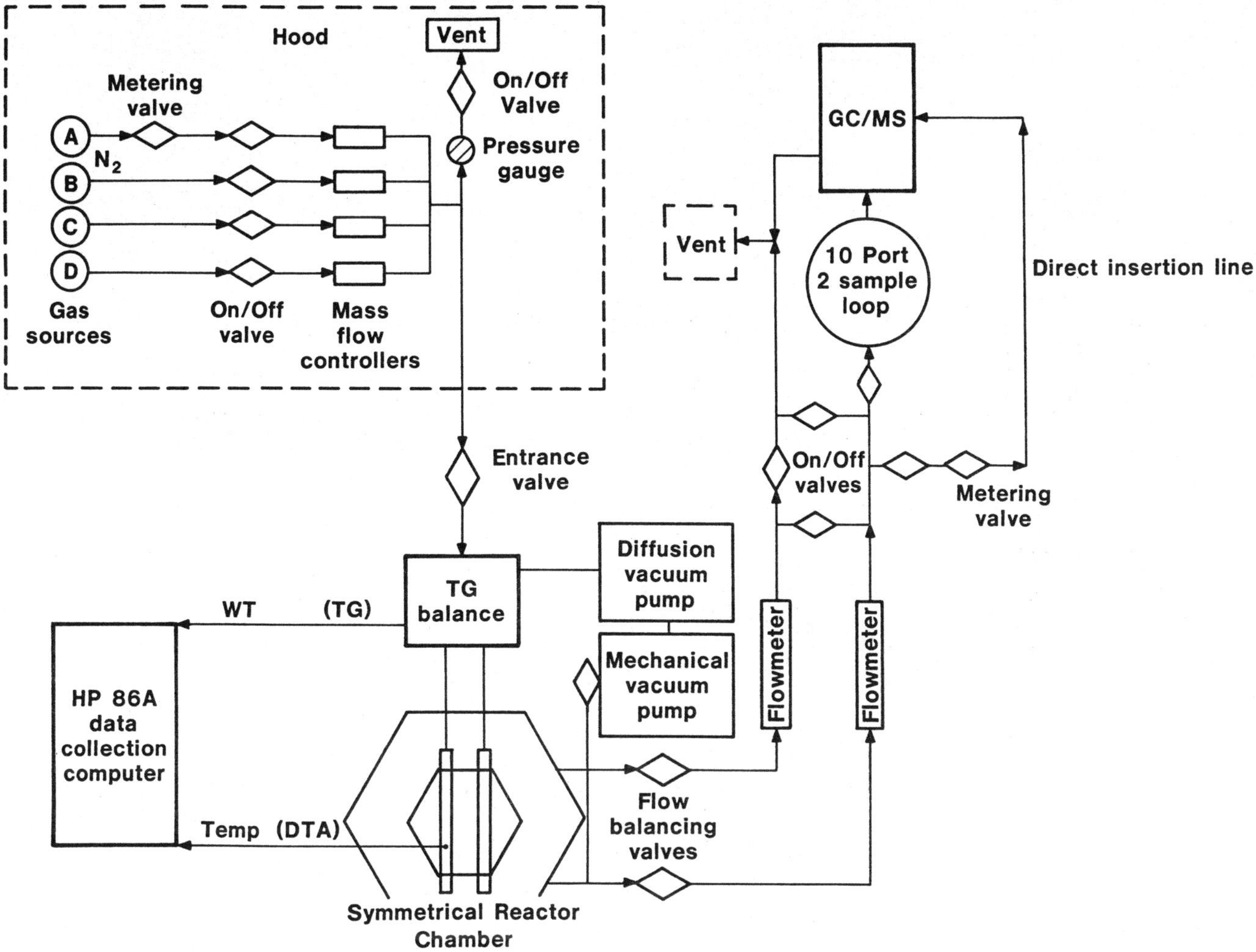

Figure 2. Microreactor system.

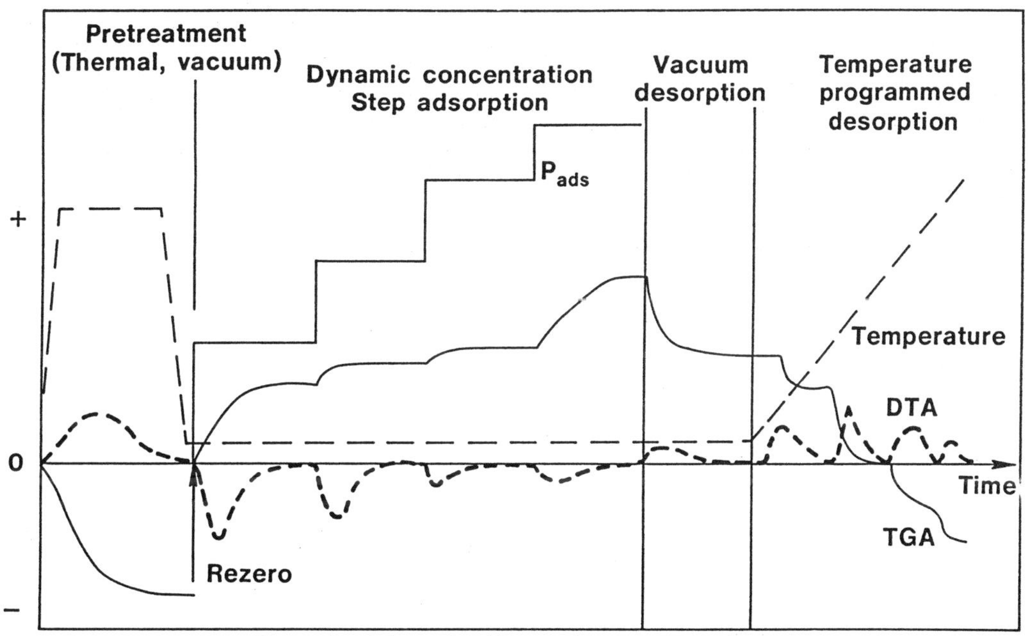

Figure 3. Adsorption/desorption run procedure.

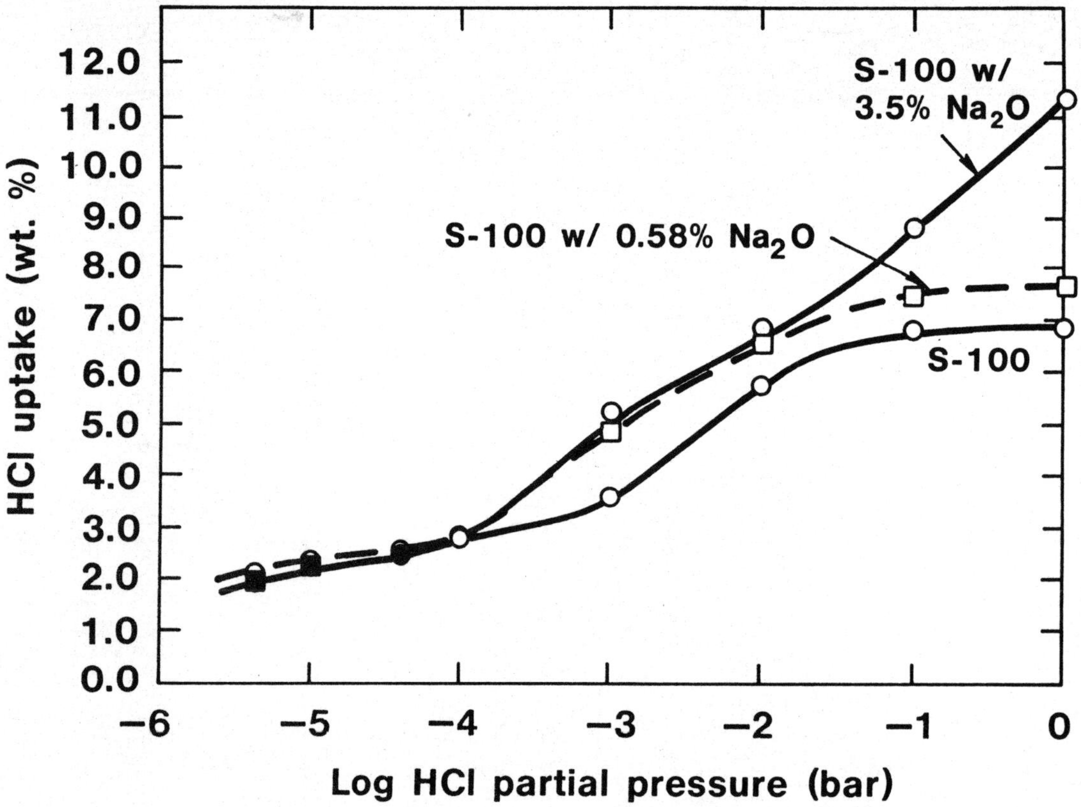

Figure 4. HCl adsorption isotherms (HCl/N_2, 25 °C, $\pi = 1$ bar, − 100 mesh particles, dynamic flow).

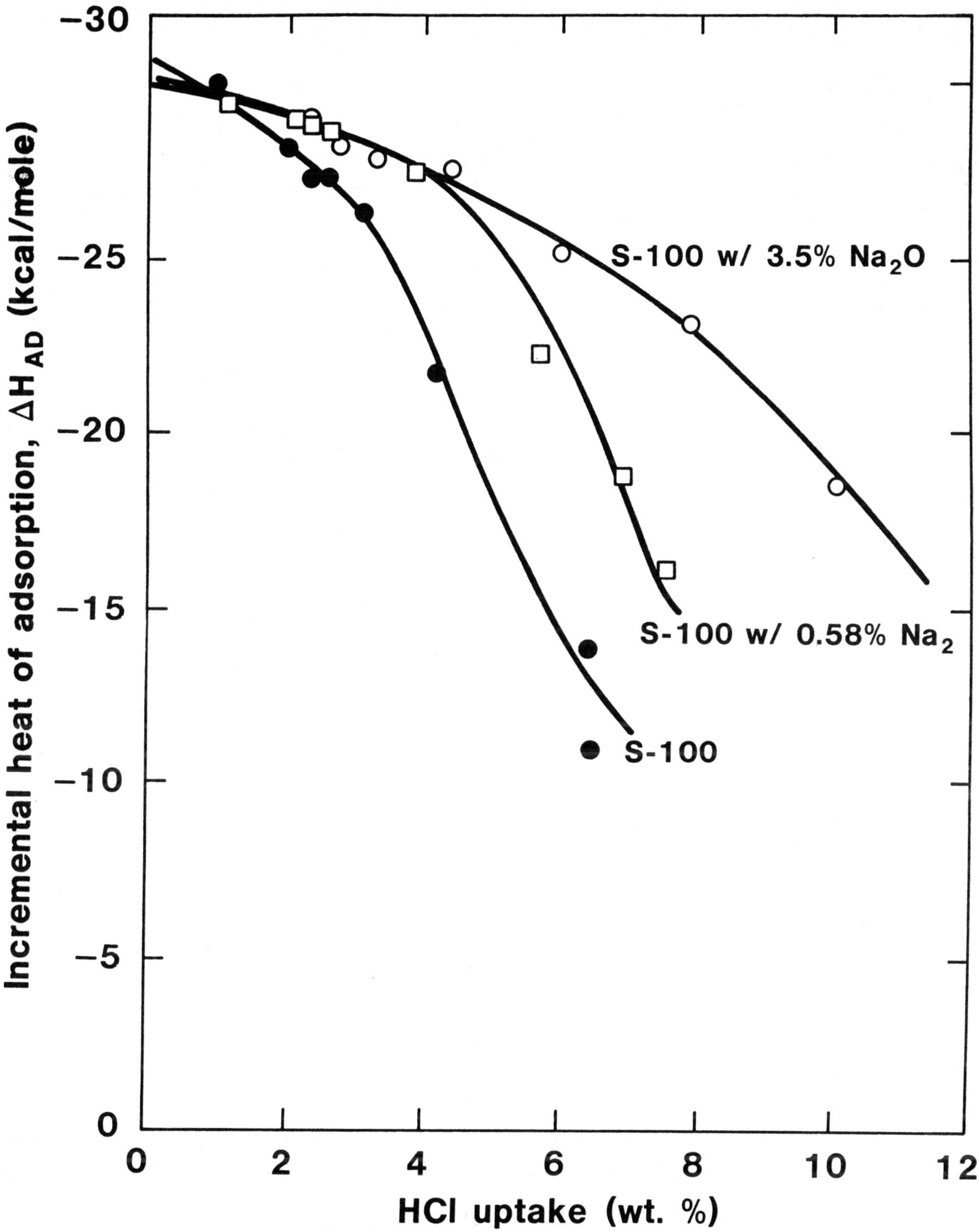

Figure 5. Differential enthalpies of HCl adsorption on aluminas, ΔH_{AD} (HCl/N_2, 25 °C, $\pi = 1$ bar, − 100 mesh particles).

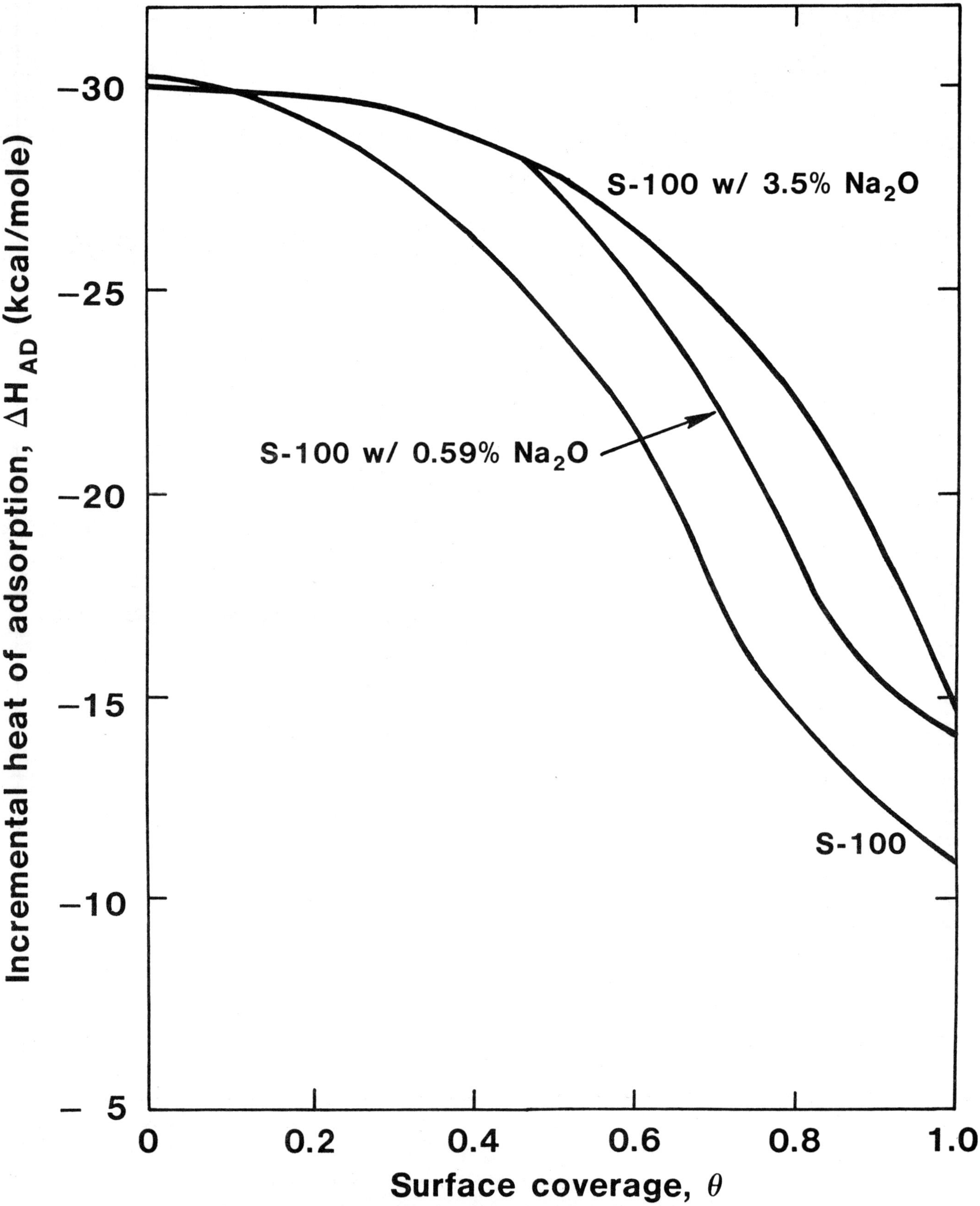

Figure 6. Energetic heterogeneity of aluminas (HCl/N_2, 25 °C, $\pi = 1$ bar, − 100 mesh particles).

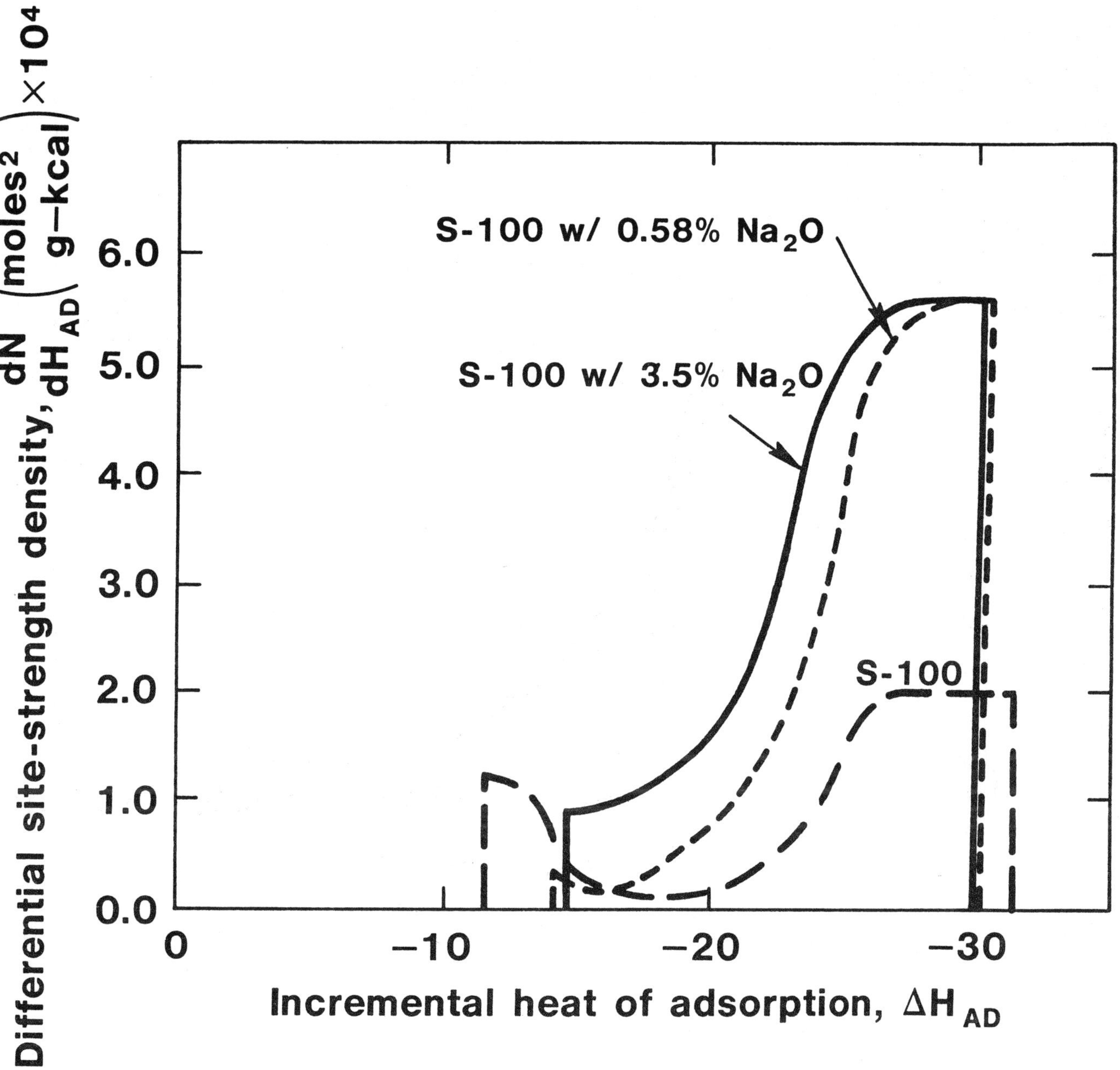

Figure 7. Site-energy for HCl on aluminas (HCl/N_2, 25 °C, $\pi = 1$ bar, − 100 mesh particles).

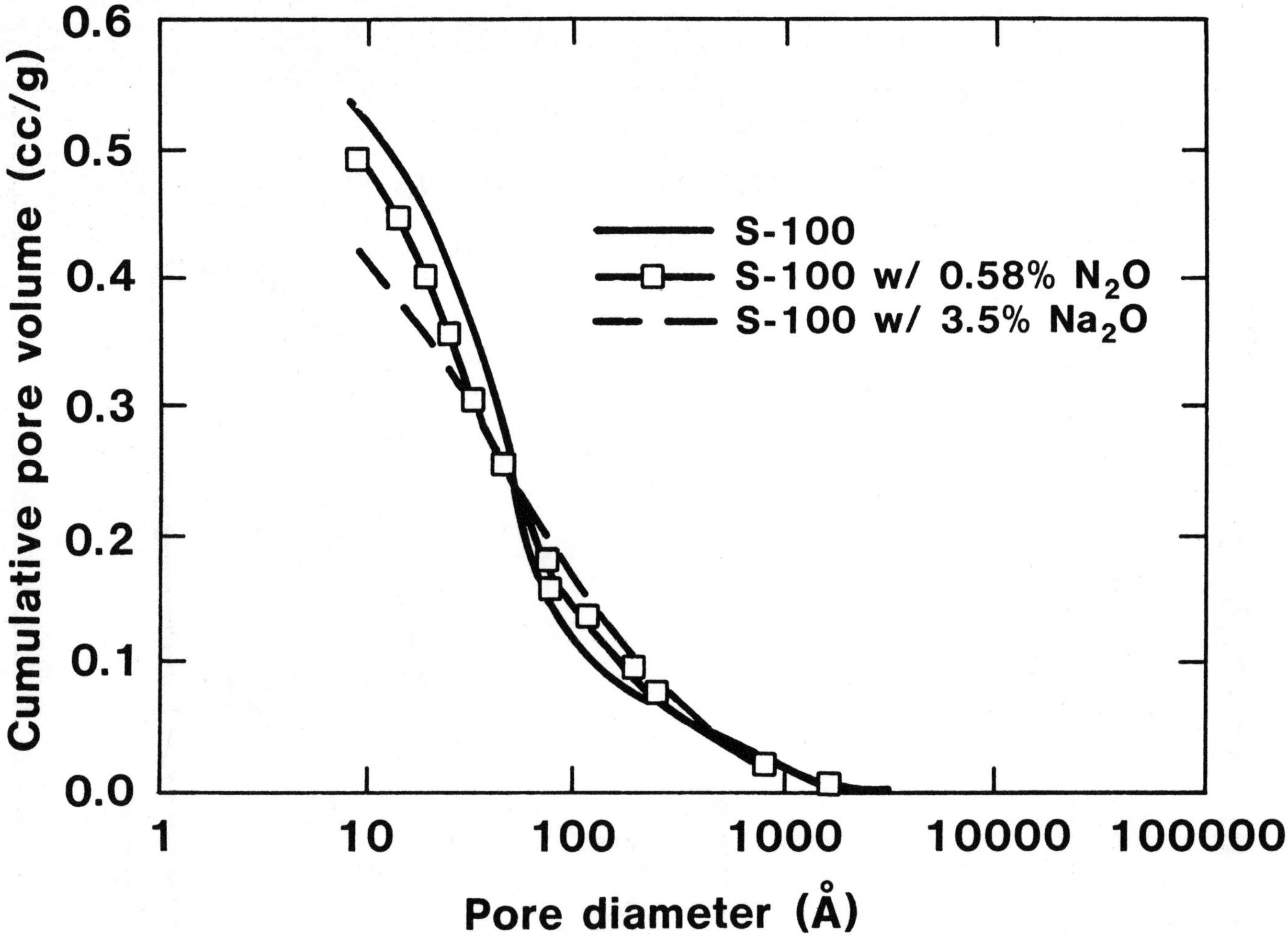

Figure 8. Pore distributions for aluminas (He and Hg porosimetry).

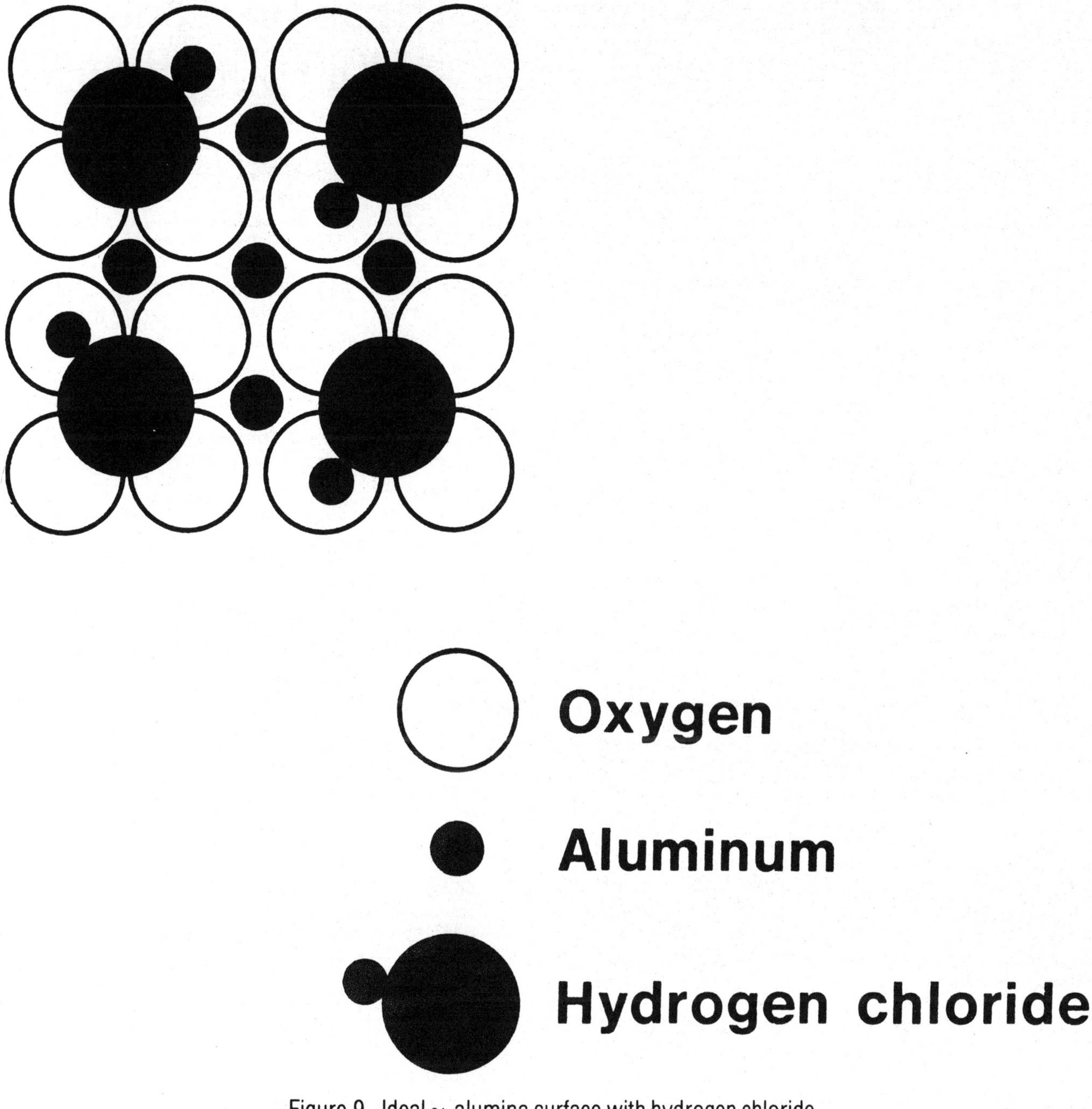

Figure 9. Ideal γ-alumina surface with hydrogen chloride adsorption.

ADSORPTION OF NITROGEN, OXYGEN, ARGON, AND TERNARY MIXTURES OF THESE GASES IN 13X MOLECULAR SIEVE

George W. Miller ■ Crew Technology Division, USAF School of Aerospace Medicine, Brooks Air Force Base, San Antonio, TX 78235-5301

Pure component and ternary adsorption equilibrium data for nitrogen, oxygen, and argon in 13X molecular sieve at temperatures of 323.15, 297.15, 233.15, and 203.15K and pressures up to 4 bar are presented. The mixture data were analyzed by a vacancy solution theory, statistical thermodynamic model, ideal adsorbed solution theory, an extended Langmuir-Sips equation, and an extended Langmuir equation. The vacancy solution theory gave the most accurate predictions of the multicomponent adsorption of nitrogen, oxygen, and argon at 323.15, 297.15 and 233.15K. In general, the predictions of all models at 203.15K deviate substantially from the mixed gas adsorption data. Surprisingly, ternary nitrogen adsorption at all temperatures closely followed the pure component isotherms. Parameters determined in this work may be useful in the design and evaluation of molecular sieve air separation systems.

Use of pressure swing adsorption for air separation has grown in popularity in recent years due to its reduced energy consumption and simplicity. Systems using this technology are usually fed compressed air at moderate pressure (~4 bar) and operated on alternating cycles of adsorption and desorption. During the adsorption phase nitrogen is preferentially adsorbed from the air feed stream resulting in enrichment of oxygen and argon. Oxygen concentrations up to 95% are easily achievable with a properly designed process (1). Although 5A and 13X molecular sieves have been successfully employed in these systems, in certain applications 13X has been shown to give better performance over a wider temperature range and to be superior to 5A at removing chemical contaminants from the feed air (2,3).

Fundamental to the design of molecular sieve air separation systems is the ability to efficiently and accurately predict mixed gas adsorption over the entire range of system temperatures and pressures. Because mixed gas adsorption data are difficult and time consuming to obtain, one generally resorts to application of an adsorption predictive model. Ideally, the model should give accurate predictions of the mixed gas adsorption based on parameters determined from analysis of pure component adsorption data, and possess a simple mathematical form to permit rapid calculation.

Although a moderate amount of work may be found in the literature concerning adsorption of nitrogen-oxygen mixtures in 5A molecular sieve, data for nitrogen-oxygen adsorption in 13X molecular sieve are scarce (4,5,6). The primary goal of this work was to investigate the applicability of a vacancy solution theory (7,8), a statistical thermodynamic model (9,10), an ideal adsorbed solution theory (11), an extended Langmuir-Sips equation, and an extended Langmuir equation to predict mixed gas adsorption of nitrogen, oxygen and argon in 13X molecular sieve over a wide pressure (up to 4 bar) and temperature range (203.15 to 323.15K). In this work the ternary adsorption predictions were based entirely on pure component parameters.

THEORY

Vacancy Solution Theory (VST)

Suwanayuen and Danner (7,8) developed a modified vacancy solution theory similar to one proposed by Dubinin (12) but incorporating modifications based on the work of Lucassen-Reynders (13,14,15). Dubinin suggested pure component adsorption on a zeolite could be viewed as an equilibrium between two vacancy solutions. Lucassen-Reynders suggested using an activity coefficient to account for non-ideality in the adsorbed phase. This theory defines a "vacancy" as a unfilled adsorption site or space, i.e. a site not yet filled by an adsorbate molecule. It is assumed the vacancies form a hypothetical solvent for the gas and adsorbed phases. Equilibrium between the two vacancy solutions is maintained by a spreading or osmotic pressure originating from the surface potential force field. A system with a gas phase consisting of a binary mixture is viewed as a ternary mixture with the third component being the solvent or vacancies. Excess properties related to a dividing surface are used to define the properties of the adsorbed phase, where the dividing surface is taken such that the partial molar area of both components (adsorbate and vacancy) are equal to a constant. This constant is the reciprocal of the limiting excess adsorption concentration (15). The pure component isotherm equation is derived by substituting the equation of state,

$$\pi = \frac{-RT}{\bar{a}_4} \ln \gamma_4^s x_4^s$$

and the Wilson equation for the activity coefficient (16),

$$\ln \gamma_4^s = -\ln(x_4^s + \Lambda_{41} x_1^s) - x_1^s \left[\frac{\Lambda_{14}}{x_1^s + \Lambda_{14} x_4^s} - \frac{\Lambda_{41}}{x_4^s + \Lambda_{41} x_1^s} \right]$$

into the Gibbs isotherm equation. The resulting pure component isotherm equation based on the modified vacancy solution theory is (7,8),

$$P = (n_1^{s,\infty}/b_1)(\theta/1-\theta) \left[\Lambda_{14} \frac{1-(1-\Lambda_{41})\theta}{\Lambda_{14}+(1-\Lambda_{14})\theta} \right]$$

$$\exp\left[-\frac{\Lambda_{41}(1-\Lambda_{41})\theta}{1-(1-\Lambda_{41})\theta} - \frac{(1-\Lambda_{14})\theta}{\Lambda_{14}+(1-\Lambda_{14})\theta} \right] \quad (1)$$

The first two terms of this expression form the Langmuir equation. The remaining terms represent the non-ideality of the system or the solvent-solute interaction in the vacancy solution. The four independent parameters $n_1^{s,\infty}$, b_1, Λ_{14}, and Λ_{41} are determined by regression analysis using the pure component data. Because the maximum adsorption, $n_1^{s,\infty}$, has been shown to remain constant at all temperatures and is most reliable if obtained at the lowest temperature, one needs only to determine three parameters at the higher temperatures. In the case of a ternary gas phase system one must solve the following equation (7,8),

$$\phi_i y_i P = \gamma_i^s x_i^s n_m^s \frac{n_i^{s,\infty} \Lambda_{i4}}{n_m^{s,\infty} b_i}$$

$$\exp(\Lambda_{41} - 1) \exp(\pi \bar{a}_i /RT) \quad (2)$$

where,

$$-\frac{\pi \bar{a}_i}{RT} = \left[1 + \frac{n_m^{s,\infty} - n_i^{s,\infty}}{n_m^s} \right] \ln \gamma_4^s x_4^s$$

$$\ln \gamma_k^s = 1 - \ln\left[\sum_{j=1}^{n} x_j^s \Lambda_{kj} \right] - \sum_{i=1}^{n} \left[x_i^s \Lambda_{ik} \Big/ \sum_{j=1}^{n} x_j^s \Lambda_{ij} \right]$$

$$n_m^{s,\infty} = x_1 n_1^{s,\infty} + x_2 n_2^{s,\infty} + x_3 n_3^{s,\infty}$$

$$x_i^s = x_i \theta$$

$$x_4^s = 1 - \theta$$

and

$$1 = y_1 + y_2 + y_3 .$$

Generally, an x_1 and x_2 are assumed and the equation is solved by trial and error. If conditions are such that the fugacity coefficients can be set equal to one, the calculation is made simpler. Also, the adsorbate-adsorbate interaction parameters may be determined by analysis of mixture data but are generally set equal to one, further simplifying the calculation. Although improvements to the VST model have been suggested (17,18), in this work only the original Wilson form of the model was employed.

Statistical Thermodynamic Model

Ruthven (9) formulated a simple theoretical adsorption isotherm for zeolites based on the statistical thermodynamic techniques found in Hill (19). The fundamental assumption of this theory states if more than one adsorbate molecule is present within a cavity the sorbate-sorbate interaction is expressed as a reduction in free cavity volume. In the application of this model two parameters, the Henry's law constant (K) and effective molecular volume (β), must be determined by analysis of pure component isotherm data. K accounts for the sorbent-sorbate interaction and is usually determined by taking the slope of the pure component isotherm at low pressure. The effective molecular volume represents the sorbate-sorbate interaction and may be determined by curve fitting the pure component data.

The pure component and binary forms of the statistical thermodynamic model have been presented elsewhere (9,10). The equation for the ternary system is,

$$n_1 = \frac{K_1 p_1 + (\Omega i)}{1 + K_1 p_1 + K_2 p_2 + K_3 p_3 + (\Omega)} \qquad (3)$$

where,

$$\Omega = \sum_i^{m_1} \sum_j^{m_2} \sum_k^{m_3} \frac{(K_1 p_1)^i \, (K_2 p_2)^j \, (K_3 p_3)^k \, (\Xi)}{i!\, j!\, k!}$$

$$\Xi = (1 - i\beta_1/V - j\beta_2/V - k\beta_3/V)^{i+j+k}$$

$$i + j + k \geq 2$$

and

$$i\beta_1 + j\beta_2 + k\beta_3 \leq V.$$

Similar expressions for n_2 and n_3 may be obtained.

Ideal Adsorbed Solution Theory (IAST)

The IAST of Myers and Prausnitz (11) suggested the adsorbates in a multicomponent mixture in thermodynamic equilibrium form an ideal solution with constant spreading pressure. The final expressions take a form analogous to Raoult's law. The isothermal Gibbs adsorption equation is applied (20),

$$A d\pi = \sum_i n_i \, d\mu_i \qquad (4)$$

The integral form of Equation 4 gives a relationship which permits calculation of the spreading pressure,

$$\pi_i^o A/RT = \int_0^{P_i^o} (n_i^o / P_i^o) \, dP_i^o \qquad (5)$$

The equations necessary for analysis of a ternary system are presented below and may be solved analytically by specifying two of the thirteen unknowns (11).

$$\pi_1^o = \Psi(P_1^o), \quad \pi_2^o = \Psi(P_2^o), \quad \pi_3^o = \Psi(P_3^o)$$

$$P y_1 = P_1^o(\pi) x_1, \quad P y_2 = P_2^o(\pi) x_2, \quad P y_3 = P_3^o(\pi) x_3$$

$$\pi_1^o = \pi_2^o = \pi_3^o$$

$$x_1 + x_2 + x_3 = 1, \quad y_1 + y_2 + y_3 = 1$$

$$n_t = [\, x_1 / n_1^o(\pi) + x_2 / n_2^o(\pi) + x_3 / n_3^o(\pi) \,]^{-1}.$$

EXPERIMENTAL

The volumetric adsorption apparatus used for measuring the pure and multicomponent adsorption equilibria is shown in Figure 1 (21,6). The volumes of the sample and charge chambers are given in Table 1. Pressure measurements were accomplished by Statham Model PA208TC-50-350 stainless steel diaphragm pressure transducers. The sample chamber was held at the desired experimental temperature by a thermostatically controlled environmental chamber with an accuracy of ±0.5K. A Model 1100 Perkin-Elmer medical gas analyzer was employed during the ternary equilibrium experiments for measurement of the gas phase composition.

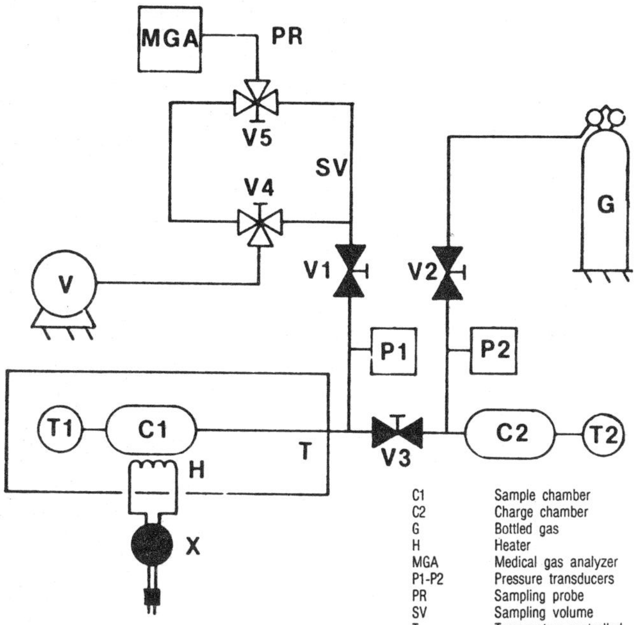

Figure 1. Adsorption apparatus.

Table 1. Dimensions of apparatus and properties of materials.

Sample chamber volume	183.8 mL
Charge chamber volume	181.8 mL
Anhydrous weight of sample	28.7 g
Particle density of anhydrous molecular sieve crystals	1.53 g/mL[a]
Solid phase density of anhydrous molecular sieve crystals and binder	3.38 g/mL[b]
Binder weight per cent	24.0 %[c]

a Reported by supplier.
b Data from this laboratory.
c Determined in this laboratory by comparing adsorption capacity of the sample with Union Carbide pure 13X molecular sieve crystals (Lot No. 9450-8406-0009).

The adsorbent employed in this work was Union Carbide, 16 x 40 mesh, 13X molecular sieve (Lot no. 9307-795-044) (see Table 1). The chemical formula for the dehydrated crystals is,

$$Na_{86}[(AlO_2)_{86}(SiO_2)_{106}]$$

Purity of the nitrogen, oxygen, and argon was 99.998%, 99.99%, and 99.98%, respectively. The concentrations of nitrogen, oxygen, argon, and carbon dioxide in the bottled air were measured using a Model 1100 Perkin-Elmer medical gas analyzer, and found to be 78.1%, 20.9%, 0.96%, and 0.03%, respectively. Water vapor content of the air was reported at ≤0.002%.

Pure component adsorption equilibria for nitrogen, oxygen, and argon were determined at temperatures of 323.15, 297.15, 233.15, and 203.15K, and up to pressures of 4 bar. Adsorbent activation was conducted at a temperature of 623K and pressure of less than 1 Torr for a period of at least 12 h. Gas from the charge chamber was allowed to expand into the sample chamber through valve V3. A minimum equilibration period of six hours was used for each pure component data point. Additional gas was added to the sample chamber until the maximum isotherm pressure (~4 bar) was achieved. Valve V1 remained closed during the pure component adsorption experiments. By applying the appropriate P-V-T measurements a mass balance was performed to determine the amount adsorbed at each step. The solid phase volume of the sample was determined by helium displacement. Compressibility correction factors were incorporated into the calculation procedure, although all gases could be assumed to behave ideally over the temperature and pressure ranges of this work. Details of the calculation procedure may be found elsewhere (21).

Multicomponent adsorption experiments involving nitrogen, oxygen, and argon were conducted at 323.15, 297.15, 233.15, and 203.15K, and up to pressures of approximately 4 bar. These experiments involved an identical procedure as in the pure component experiments except air was injected in an identical fashion until the maximum equilibrium pressure was attained. Hence, the total molar ratio within the sample chamber remained constant. After equilibrium was attained a small gas sample was removed from the sample chamber through valve V1 for concentration analysis. The mass of each of the three components in the sample chamber was then corrected for the amount removed during the sampling process. Throughout the pure component and multicomponent experimental work two isotherms were accomplished at each temperature to ensure the reproducibility of the data.

Attainment of true equilibrium in the multicomponent adsorption experiments was assured by allowing a sufficiently long equilibration period for each data point and by checking the data for thermodynamic consistency. Due to the static nature of the multicomponent adsorption experiments one must ensure the adsorbent and gas mixture have attained true equilibrium. Because gas phase mixing will occur at the slowest rate at the lowest temperature (203.15K), preliminary isotherms were collected at different equilibration periods and 203.15K. It was observed that equilibration periods >10 h resulted in a negligible change in the amount of each component adsorbed. Hence, each multicomponent data point in this work was permitted an equilibration period of at least 10 h. Additionally, the multicomponent data were checked for thermodynamic consistency by application of the Gibbs adsorption isotherm (Equation 4).

RESULTS AND DISCUSSION

Pure Component Adsorption

The pure component isotherms of nitrogen, oxygen, and argon on 13X molecular sieve at temperatures of 323.15, 297.15, 233.15, and 203.15K are presented in Figure 2. The data are tabulated in Table 2. The oxygen and argon data were best correlated with a Langmuir equation while the nitrogen data were best correlated with a Sips equation (22):

$$n_{O_2 \text{ or } Ar} = \frac{a\,P}{1 + b\,P} \; ; \; n_{N_2} = \frac{a\,P^c}{1 + b\,P^c}$$

The values for the constants were determined by application of a least squares regression analysis and are given in Table 3. The amount adsorbed, n, has units of molecules/cavity and pressure, P, has units of bar. In subsequent calculations these equations were used to represent the data.

Heat of Adsorption

The isosteric heat of adsorption for the pure gases was determined by application of the Clausius-Clapeyron equation with assumptions that the ideal gas law applies and the volume of the adsorbed phase is negligible when compared to the gas phase volume. The results of this analysis are shown in Figure 3. At low coverage the isosteric heat of adsorption for nitrogen on 13X molecular sieve is considerably less than that observed for 5A molecular sieve (6). These higher values at low coverage are attributed to the quadrupole moment of the nitrogen molecule interacting with the highest energy sites in the molecular sieve crystal. From Figure 3 it appears at a coverage of two molecules/cavity these sites are filled. For both oxygen and argon, the isosteric heat of adsorption was nearly independent of coverage.

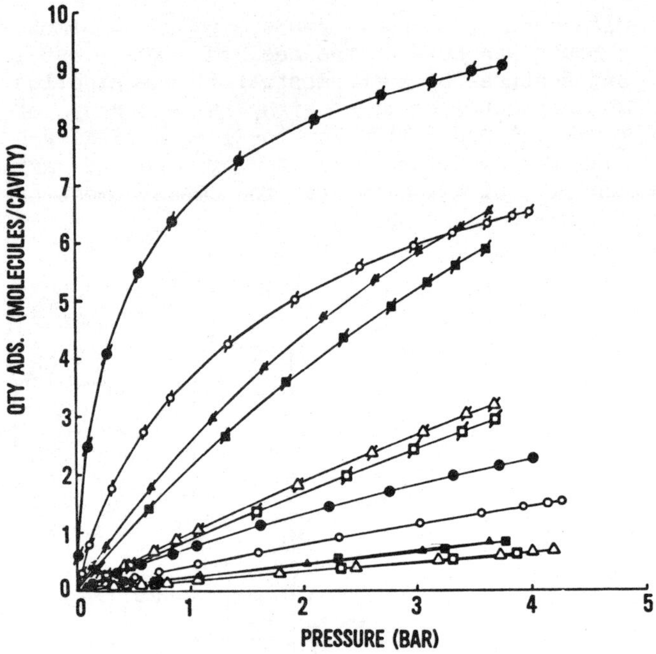

Figure 2. Pure component equilibrium data of nitrogen, oxygen, and argon in 13X molecular sieve.

	323.15K	297.15K	233.15K	203.15K
Nitrogen	○	●	○ (flagged)	● (flagged)
Oxygen	△	▲	△ (flagged)	▲ (flagged)
Argon	□	■	□ (flagged)	■ (flagged)

Prediction of Pure Component Isotherms

The statistical thermodynamic model was previously shown useful in estimating pure component isotherms of nitrogen, oxygen, and argon in 5A molecular sieve (6). Equation 3 simplified to the pure component expression may be used to estimate pure component isotherms over a wide range of temperatures and pressures if valid values for K and β are known. Here K values for the oxygen-13X and argon-13X data were found by taking the derivative of the Langmuir equation at zero

Table 2. Pure component data for oxygen, nitrogen, and argon in 13X molecular sieve.

Gas	Temp.(K)							
	323.15K		297.15K		233.15K		203.15K	
	P	n	P	n	P	n	P	n
O_2	0.272	0.050	0.139	0.041	0.191	0.199	0.147	0.402
	0.537	0.094	0.447	0.117	0.413	0.420	0.285	0.764
	0.763	0.132	0.699	0.178	0.664	0.674	0.692	1.804
	1.052	0.187	1.069	0.273	0.855	0.861	1.220	2.941
	1.778	0.286	2.033	0.472	1.041	1.050	1.668	3.849
	2.447	0.390	3.040	0.705	1.930	1.807	2.164	4.716
	3.171	0.511	3.633	0.837	2.585	2.343	2.612	5.369
	3.722	0.598			3.040	2.712	3.040	5.814
	3.991	0.644			3.426	3.006	3.391	6.173
	4.184	0.684			3.674	3.193	3.653	6.460
N_2	0.261	0.119	0.141	0.120	0.043	0.317	0.016	0.617
	0.509	0.227	0.364	0.296	0.113	0.797	0.107	2.484
	0.720	0.317	0.573	0.455	0.301	1.758	0.276	4.078
	1.053	0.455	0.852	0.658	0.588	2.747	0.568	5.484
	1.592	0.639	1.045	0.790	0.823	3.343	0.873	6.364
	2.295	0.895	1.627	1.113	1.351	4.224	1.461	7.346
	3.019	1.137	2.226	1.443	1.937	4.969	2.123	8.035
	3.557	1.313	2.757	1.712	2.509	5.514	2.709	8.457
	3.922	1.422	3.336	1.973	2.978	5.879	3.157	8.733
	4.136	1.481	3.736	2.137	3.322	6.120	3.508	8.904
	4.267	1.521	4.026	2.264	3.626	6.315	3.777	9.014
					3.846	6.435		
					3.991	6.517		
Ar	0.408	0.071	0.424	0.106	0.245	0.234	0.168	0.377
	0.696	0.118	0.692	0.169	0.457	0.422	0.644	1.409
	1.043	0.178	2.295	0.507	1.578	1.338	1.323	2.649
	2.330	0.361	3.246	0.710	2.364	1.979	1.861	3.577
	3.302	0.512	3.784	0.830	2.964	2.411	2.357	4.339
	3.874	0.607			3.384	2.716	2.764	4.864
					3.681	2.924	3.102	5.282
							3.357	5.564

pressure. K values for the nitrogen data and β values for all the pure component gases were determined by curve fitting the pure component isotherm data to the statistical thermodynamic equation. These derived K values were fitted to a linearized form of the vant Hoff equation $\{K = K_o \exp(q_o/RT)\}$ by a least squares technique. The results of this analysis are shown in Tables 4 and 5. Although the values for the limiting heat of adsorption for oxygen and argon agree closely with those shown in Figure 3, the value for nitrogen differs slightly. The effective molecular volumes for oxygen, argon, and nitrogen remained nearly constant over the entire temperature range. Also, β values for oxygen and argon in this work agree with those determined previously (6).

By application of the vant Hoff equation and the parameters derived in this work one may predict isotherms of nitrogen, oxygen, and argon in 13X molecular sieve at other temperatures and over a pressure range of 0-4 bar. Using K values calculated from the vant Hoff equation, and β values given in Table 5, one applies the pure component form of the statistical thermodynamic model. In Figure 4 the predictions of nitrogen-13X isotherms at 283, 298, and 313K are in good agreement with the data of Peterson (23).

Table 3. Constants for the pure component Langmuir and Sips equations.

Gas	Temp.(K)	a	b	c
O_2	323.15	0.166	0.0068	——
	297.15	0.254	0.0295	——
	233.15	1.046	0.0565	——
	203.15	3.038	0.1934	——
Ar	323.15	0.166	0.0177	——
	297.15	0.234	0.0202	——
	233.15	0.913	0.0406	——
	203.15	2.382	0.1298	——
N_2	323.15	0.441	0.0369	0.948
	297.15	0.812	0.0889	0.939
	233.15	5.777	0.5796	0.843
	203.15	15.893	1.3904	0.729

n and P have units of molecules/cavity and bar, respectively.

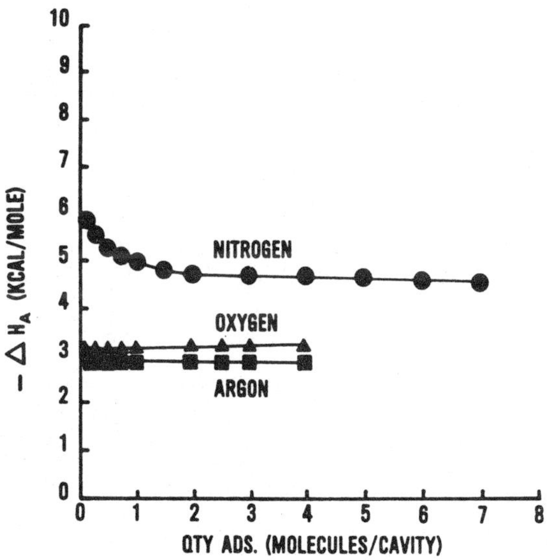

Figure 3. Isosteric heat of adsorption for nitrogen, oxygen, and argon in 13X molecular sieve.

Multicomponent Adsorption

Multicomponent adsorption data of N_2-O_2-Ar-13X taken at a constant total molar ratio of 0.7813:0.2091:0.0096, and temperatures of 323.15, 297.15, 233.15, and 203.15K are shown in Figures 5-9, and tabulated in Table 6. Surprisingly, the ternary nitrogen adsorption data closely follows the nitrogen pure component isotherms at all temperatures. The adsorbed phase mole fractions at each temperature were nearly constant. The low concentrations of carbon dioxide and water in the bottled air were assumed to have a negligible effect upon the mixed gas adsorption data.

Table 4. Parameters K_o and q_o for the vant Hoff equation. $\{K = K_o \exp(q_o/RT)\}$.

System	K_o (molecules/ cavity bar)	q_o (kcal/mole)
O_2-13X	1.23×10^{-3}	3.14
Ar-13X	1.74×10^{-3}	2.91
N_2	4.41×10^{-4}	4.51

Table 5. K values and effective molecular volumes based on the pure component isotherms.

System	Temp. (K)	K (molecules/ cavity·bar)	m	β (Å^3)
O_2-13X	323.15	0.166	32	35.2
	297.15	0.254	32	35.2
	233.15	1.046	32	35.2
	203.15	3.038	32	35.2
Ar-13X	323.15	0.166	38	29.8
	297.15	0.235	38	29.8
	233.15	0.913	38	29.8
	203.15	2.382	38	29.8
N_2-13X	323.15	0.49	14	82.1
	297.15	0.90	14	82.1
	233.15	8.0	14	82.1
	203.15	30.0	15	76.6

V = 1150 Å^3 (25).
One molecule/cavity is equivalent to 0.556 mmole/gm of anhydrous molecular sieve crystals or 0.423 mmole/gm of anhydrous molecular sieve pellets with 24% inert binder by weight.

The vacancy solution theory gave excellent predictions of the mixed gas adsorption of nitrogen, oxygen, and argon at 323.15, 297.15, and 233.15K (Figures 5-7 and 9). In general, at 203.15K all models gave rather poor predictions of the mixed gas adsorption. The calculation proceeds by determining values for $n_i^{s,\infty}$, b_i, Λ_{i4}, and Λ_{4i} by regression analysis of the pure component data using Equation 1. The results of this analysis are shown in Table 7. The maximum loading of component i, $n_i^{s,\infty}$, was determined at 203.15K where the value obtained is more accurate, and hence, this value was employed in the subsequent regression procedures at the higher

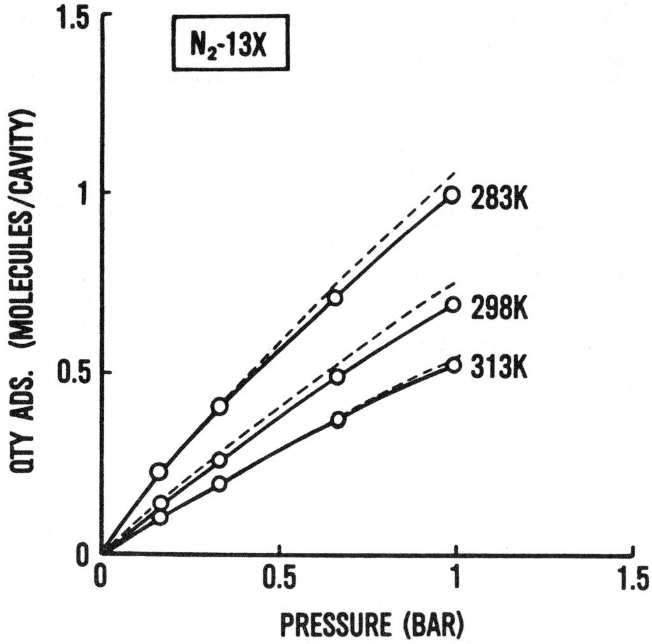

Figure 4. Prediction of the pure component data of Peterson (*23*) using the statistical thermodynamic model and parameters derived from this work. (--- , prediction)

temperatures. Binary (N_2-O_2) and ternary (N_2-O_2-Ar) multicomponent adsorption predictions were accomplished using Equation 2 with assumptions that the fugacity coefficients may be set equal to one and the adsorbate-adsorbate interactions are negligible. Both the binary and ternary form of the model gave nearly identical results for nitrogen and oxygen adsorption. More accurate predictions at 203.15K may have been possible had the adsorbate-adsorbate interaction parameters been included in the calculation.

The binary form of the statistical thermodynamic model was used to determine the mixed gas adsorption of nitrogen and oxygen (Figures 5-8). The parameters in the model were taken from Table 5. This model gave reasonable predictions of the nitrogen-oxygen adsorption at all temperatures except 203.15K. It has been previously shown for adsorption on 5A molecular sieve that elimination of argon from the model has a negligible effect on the binary N_2-O_2 adsorption at the conditions of this work (6). Although the calculation procedure was rapid for the binary model, extending the model to a ternary system resulted in significantly longer computer calculation times making the ternary form of the model impractical for 13X ternary mixture predictions for the N_2-O_2-Ar system.

Table 6. Gas mixture adsorption data for nitrogen, oxygen, and argon in 13X molecular sieve.

Data Point	p_{N_2}	p_{O_2}	p_{Ar}	n_{N_2}	n_{O_2}	n_{Ar}
			323.15K			
A	0.212	0.082	0.0036	0.111	0.016	0.0008
B	0.318	0.120	0.0054	0.159	0.023	0.0011
C	0.484	0.181	0.0079	0.237	0.034	0.0017
D	0.725	0.265	0.0113	0.347	0.051	0.0026
E	1.743	0.622	0.0265	0.723	0.104	0.0054
F	2.366	0.818	0.0357	0.946	0.142	0.0070
			297.15K			
G	0.212	0.094	0.0043	0.191	0.029	0.0013
H	0.375	0.164	0.0074	0.320	0.047	0.0021
I	0.706	0.306	0.0136	0.579	0.083	0.0038
J	1.523	0.634	0.0286	1.070	0.146	0.0065
K	2.079	0.794	0.0361	1.352	0.206	0.0090
L	2.445	0.906	0.0407	1.525	0.235	0.0105
			233.15K			
M	0.238	0.260	0.0126	1.535	0.265	0.0113
N	0.592	0.529	0.0240	2.670	0.432	0.0192
O	0.978	0.718	0.0337	3.467	0.567	0.0243
P	1.344	0.877	0.0400	4.020	0.650	0.0286
Q	1.670	1.000	0.0459	4.415	0.708	0.0309
			203.15K			
R	0.061	0.130	0.0066	2.439	0.557	0.0244
S	0.202	0.299	0.0148	3.990	0.855	0.0372
T	0.409	0.491	0.0245	5.250	1.066	0.0457
U	0.751	0.684	0.0332	6.167	1.209	0.0520
V	1.059	0.890	0.0428	6.855	1.267	0.0542
W	1.389	1.029	0.0494	7.315	1.322	0.0563
X	1.659	1.147	0.0549	7.622	1.337	0.0566

The ternary form of the ideal adsorbed solution theory was applied to the mixed gas adsorption data and gave reasonable predictions at all temperatures except 203.15K (Figures 5-9). Spreading pressures were calculated by numerical integration of Equation 5, and isotherm curves fitted to a Sips equation for nitrogen and a Langmuir equation for oxygen and argon. The numerical integration was accomplished using pressure increments of 0.01 bar. The results for nitrogen and oxygen indicate the predictions of this model closely follow those of the statistical thermodynamic model at most temperatures.

A ternary form of an extended Langmuir-Sips equation using constants given in Table 3 was used to predict the ternary mixture data (Figures 5-9). The nitrogen form of this equation is,

$$n_{N_2} = \frac{a_{N_2} p_{N_2}^{c}}{1 + b_{Ar} p_{Ar} + b_{O_2} p_{O_2} + b_{N_2} p_{N_2}^{c}}$$

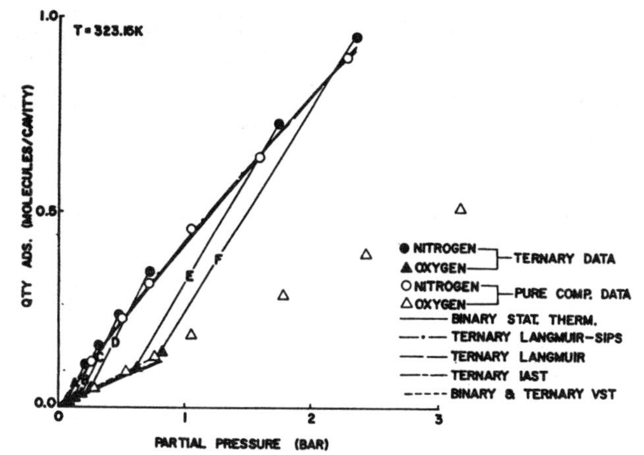

Figure 5. Prediction of mixed gas adsorption in 13X molecular sieve at 323.15K. Data points connected by lines A-F and tabulated in Table 6.

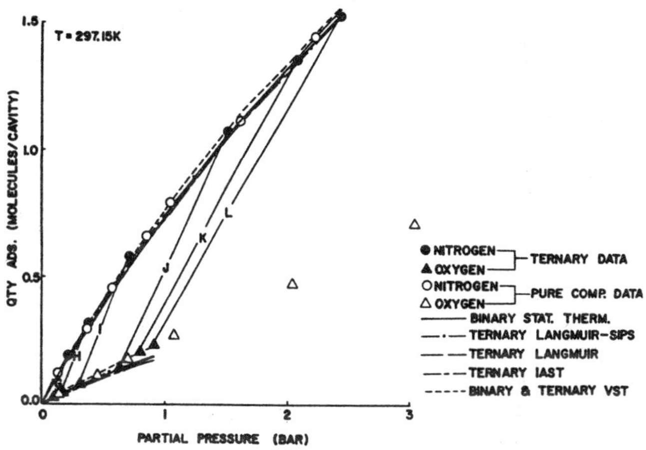

Figure 6. Prediction of mixed gas adsorption in 13X molecular sieve at 297.15K. Data points connected by lines G-L and tabulated in Table 6.

This equation gave good predictions of the mixed gas adsorption data at most temperatures. The mathematical simplicity of this model is a definite advantage.

Prediction of the mixture data was also attempted by an extended Langmuir equation of the form (Figures 5-9),

$$n_{N_2} = \frac{r_{N_2} p_{N_2}}{1 + b_{Ar} p_{Ar} + b_{O_2} p_{O_2} + s_{N_2} p_{N_2}}$$

The constants for oxygen and argon were taken from Table 3. The nitrogen constants were determined by a regression analysis procedure identical to that used for oxygen and argon. Generally, this equation gave the least accurate predictions of all the predictive methods in this work.

Table 7. Regression parameters of the vacancy solution model for adsorption of oxygen, nitrogen, and argon in 13X molecular sieve.

Gas	$n_i^{s,\infty}$ (molecule/cavity)	Temp. (K)	b_i (molecule/cavity·bar)	Λ_{i4}	Λ_{4i}	Sum of Squares of Relative Residuals
O_2	10.766	323.15	0.166	1.256	1.64×10^{-8}	0.0191
		297.15	0.252	2.499	2.95×10^{-7}	0.0077
		233.15	1.045	1.416	9.58×10^{-9}	0.0014
		203.15	2.676	0.787	0.826	0.0034
N_2	12.264	323.15	0.452	0.506	1.975	0.0027
		297.15	0.865	0.447	2.237	0.0019
		233.15	10.139	1.159	1.819	0.0004
		203.15	43.172	1.914	1.411	0.0007
Ar	8.556	323.15	0.165	1.715	1.65×10^{-7}	0.0109
		297.15	0.234	1.531	2.68×10^{-8}	0.0050
		233.15	0.913	1.241	6.21×10^{-9}	0.0009
		203.15	2.287	1.193	0.191	0.0051

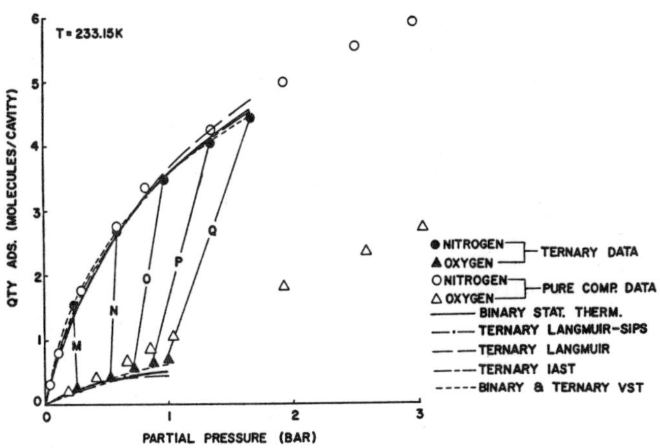

Figure 7. Prediction of mixed gas adsorption in 13X molecular sieve at 233.15K. Data points connected by lines M-Q and tabulated in Table 6.

Thermodynamic Consistency

A thermodynamic consistency test of the mixture data was accomplished through application of the Gibbs adsorption isotherm equation (24). The fugacities in the gas and adsorbed phases are equal at equilibrium. Hence,

$$P y_i \phi_i = P_i^o(\pi)\, \phi_i^o \gamma_i x_i \qquad (6)$$

may be written for each component. The gas phase fugacity coefficients were assumed equal to one. From the integrated form of Equation 4 the spreading pressure of the mixture was determined:

$$\pi A/RT = \int_0^{p_1} (n_1/p_1)\,dp_1 + \int_0^{p_2} (n_2/p_2)\,dp_2 + \int_0^{p_3} (n_3/p_3)\,dp_3$$

$$1, 2, 3 = N_2, O_2, Ar$$

Activity coefficients were calculated using Equation 6 and the experimental data. The excess Gibbs free energy was calculated,

$$\Delta g^e/RT = \Sigma\, x_i \ln \gamma_i$$

The excess reciprocal loading was determined for each data point:

$$(1/n)^e = (1/n) - (1/n)^{id}$$

where,

$$(1/n)^{id} = \Sigma\, (x_i / n_i^o)$$

The results of these calculations are shown in Table 8. It can be shown from Equation 4 that,

$$(1/n)^e = \left[\frac{\partial(\Delta g^e/RT)}{\partial(\pi A/RT)}\right]_{T,x} \qquad (7)$$

The slope of the excess Gibbs free energy versus the spreading pressure for each data point agreed closely with the values obtained for the reciprocal excess loading (differences of $\leq 10\%$). This agreement indicates a high degree of thermodynamic consistency.

CONCLUSIONS

In general, the vacancy solution theory gave the most accurate predictions of the mixed gas adsorption of nitrogen, oxygen, and argon in 13X molecular sieve. The extended

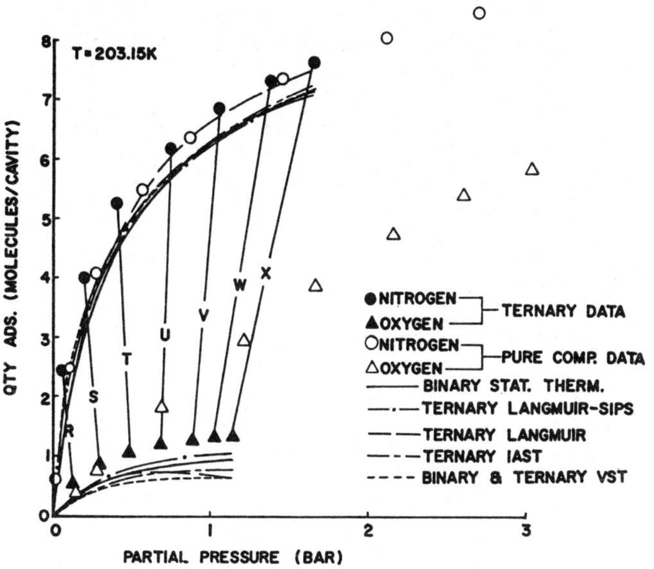

Figure 8. Prediction of mixed gas adsorption in 13X molecular sieve at 203.15K. Data points connected by lines R-X and tabulated in Table 6.

Table 8. Results of the thermodynamic consistency test.

Data Pt.	πA/RT (molecule/cavity)	γ N_2	γ O_2	γ Ar	$(\Delta g^e/RT)$	$(1/n)^e$ (molecule/cavity)$^{-1}$	RHS of Eqn. 7	Diff. (RHS -LHS) of Eqn. 7	% Diff.
					323.15K				
A	0.128	0.950	0.835	0.740	-0.0690	-0.4029	-0.4167	-0.0138	-3.4
B	0.189	0.937	0.821	0.780	-0.0830	-0.1185	-0.1140	0.0045	3.9
C	0.284	0.925	0.849	0.726	-0.0906	-0.0496	-0.0539	-0.0043	-8.3
D	0.417	0.921	0.827	0.676	-0.0980	-0.0509	-0.0515	-0.0006	-1.2
E	0.935	0.923	0.867	0.688	-0.0900	0.0428	0.0462	0.0034	7.6
F	1.219	0.941	0.835	0.711	-0.0779	0.0139	0.0140	0.0001	0.7
					297.15K				
G	0.222	1.035	0.806	0.768	-0.0003	-0.3080	-0.3030	0.0050	1.6
H	0.387	0.997	0.821	0.775	-0.0296	-0.0731	-0.0769	-0.0038	-5.1
I	0.705	0.969	0.845	0.769	-0.0498	-0.0503	-0.0471	0.0032	6.6
J	1.404	0.960	0.881	0.846	-0.0520	0.0088	0.0085	-0.0003	-3.5
K	1.821	0.984	0.756	0.747	-0.0523	0.0019	0.0021	0.0002	10.0
L	2.085	0.989	0.734	0.704	-0.0523	0.0014	0.0013	-0.0001	-7.4
					233.15K				
M	2.019	1.075	0.872	0.873	0.0407	-0.0682	-0.0675	0.0007	1.0
N	4.184	0.988	0.851	0.775	-0.0341	-0.00407	-0.0043	-0.0002	-4.8
O	5.888	0.998	0.775	0.762	-0.0391	-0.00022	-0.00021	0.00001	4.2
P	7.197	1.007	0.752	0.706	-0.0356	0.00101	0.00111	0.00001	0.9
Q	8.206	1.014	0.734	0.703	-0.0326	0.00108	0.00117	0.00009	8.0
					203.15K				
R	3.003	0.983	0.644	0.596	-0.0989	-0.1393	-0.1393	0	0
S	7.801	0.673	0.514	0.476	-0.4464	-0.00307	-0.00315	-0.00008	-2.6
T	11.669	0.650	0.512	0.496	-0.4731	0.00058	0.00063	0.00005	7.5
U	15.649	0.659	0.467	0.448	-0.4754	0.00353	0.00357	0.00004	1.1
V	18.234	0.657	0.486	0.474	-0.4690	0.00138	0.00125	-0.00013	-9.7
W	20.390	0.663	0.458	0.457	-0.4701	0.00062	0.00063	0.00001	0.8
X	21.869	0.662	0.449	0.457	-0.4729	0.00012	0.00011	-0.00001	-8.7

Data points correspond to those tabulated in Table 6.

Langmuir equation gave the least accurate predictions. At 203.15K all the models underestimate the amount adsorbed of each component. At the higher temperatures (323.15 and 297.15K) all models gave reasonable predictions of the mixed gas adsorption. Improvement in the vacancy solution theory predictions at 203.15K may have been possible had the

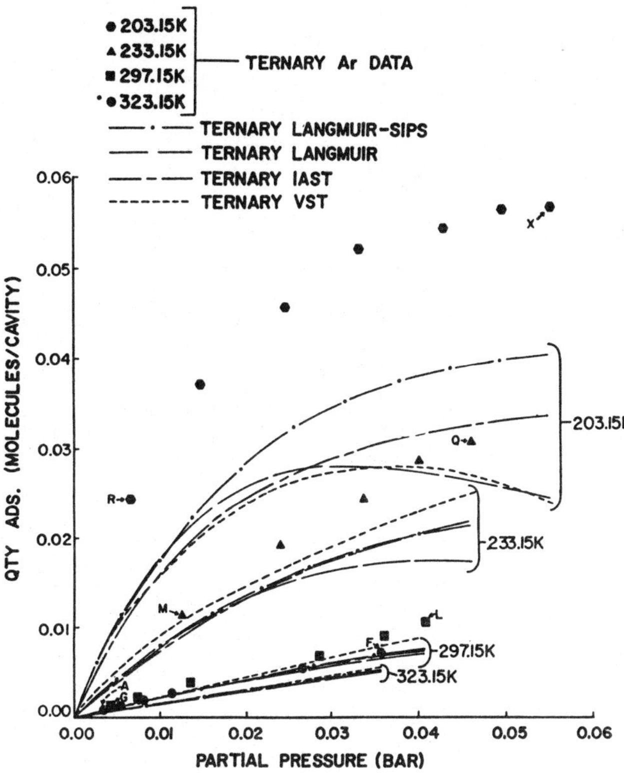

Figure 9. Prediction of ternary argon adsorption in 13X molecular sieve. Data points A-X shown in Table 6.

adsorbate-adsorbate interaction parameters been included in the model.

The statistical thermodynamic model and IAST gave reasonable predictions of the mixed gas adsorption at most temperatures. Also, the pure component form of the statistical thermodynamic model should provide a convenient method for estimating pure component isotherms of nitrogen, oxygen, and argon in 13X molecular sieve. The extended Langmuir-Sips equation gave good predictions at most temperatures with the added advantage of mathematical simplicity. All models permitted rapid calculation except the ternary form of the statistical thermodynamic model.

ACKNOWLEDGEMENTS

The author gratefully acknowledges the assistance of Daniel Sears, TSgt Victor Elizondo, A1C Robert McElvoy, Mike Briere, Clarence Theis, and Marilyn McConkey of the USAF School of Aerospace Medicine, in the preparation of this manuscript.

NOMENCLATURE

A	specific area of adsorbent
a,b,c,r,s	constants in the Langmuir and Sips equations
$\bar{a}_i$	partial molar surface area of i
b_i	Henry's Law constant of component i for the vacancy solution theory
Δg^e	excess Gibbs free energy function
H	isosteric heat of adsorption (Kcal/mol)
i,j,k	integers
K	Henry's Law constant for the statistical thermodynamic model (molecules/cavity bar)
K_o, q_o	parameters of the vant Hoff equation
m	saturation limit (V/β) (molecules/cavity)
n	amount adsorbed (molecules/cavity)
$n_i^o(\pi)$	amount adsorbed of pure component i at the spreading pressure of the mixture
P	total pressure (bar)
p	partial pressure (bar)
$P_i^o(\pi)$	equilibrium pressure of pure component i at the spreading pressure of the mixture
R	universal gas constant
T	temperature (K)
V	cavity volume for 13X molecular sieve (1150 Å^3)
x	mole fraction in adsorbed phase
y	mole fraction in gas phase

Greek symbols

β	effective molecular volume (Å^3)
γ	activity coefficient
π	spreading pressure of adsorbed phase
θ	fractional coverage
$\Lambda_{ij}, \Lambda_{ji}$	Wilson parameters for interaction of i and j
μ	chemical potential
φ	gas phase fugacity coefficient
Ψ	function

Superscripts

e	excess function
id	ideal value
o	standard state
s	surface phase value
∞	value at saturation

Subscripts

1,2,3	components 1, 2, and 3
4	vacancy
i,j,k	components i, j, and k
m	value for the mixture
n	total number of components in a mixture
t	total

LITERATURE CITED

1. Lee, H. and D. E. Stahl, AIChE Symp., Ser., 134, 69, 1 (1973).
2. Miller, G. W., K. G. Ikels, and P. A. Lozano, SAFE J., 16, 4, 28 (1986).
3. Miller, G. W., and C. F. Theis, "Performance Studies on a Molecular Sieve Oxygen Concentrator (MSOC): Comparison of MG3, 5AMG, and 13X Molecular Sieves," presented at the 24th Annual SAFE (formerly Survival and Flight Equipment) Symposium, San Antonio, Texas (1986).
4. Ruthven, D. M., AIChE J., 22, 4, 753 (1976).

5. Sorial, G. A., W. H. Granville, and W. O. Daly, Chem. Eng. Sci., 38, 9, 1517 (1983).

6. Miller, G. W., K. S. Knaebel, and K. G. Ikels, AIChE J., 33, 2, 194 (1987).

7. Suwanayuen, S., and R. P. Danner, AIChE J., 26, 1, 68 (1980).

8. Suwanayuen, S., and R. P. Danner, AIChE J., 26, 1, 76 (1980).

9. Ruthven, D. M., Nat. Phys. Sci., 232, 70 (1971).

10. Ruthven, D. M., K. F. Loughlin, and K. A. Holborrow, Chem. Eng. Sci., 28, 701 (1973).

11. Myers, A. L., and J. M. Prausnitz, AIChE J., 11, 121 (1965).

12. Dubinin, M. M., "New Results in Investigation of Equilibria and Kinetics of Adsorption of Gases on Zeolites," presented at the Fourth International Conference on Molecular Sieves, University of Chicago, (1977).

13. Lucassen-Reynders, E. H., J. Colloid Interface Sci., 41, 156 (1972).

14. Lucassen-Reynders, E. H., J. Colloid Interface Sci., 42, 554 (1973).

15. Lucassen-Reynders, E. H., Prog. Surf. Membrane Sci., 10, 253 (1976).

16. Wilson, G. M., J. Am. Chem. Soc., 86, 127 (1964).

17. Cochran, T. W., R. L. Kabel, and R. P. Danner, AIChE J., 31, 2, 268 (1985).

18. Cochran, T. W., R. L. Kabel, and R. P. Danner, AIChE J., 31, 12, 2075 (1985).

19. Hill, T. L., Introduction to Statistical Thermodynamics, Addison-Wesley, Reading, Mass. (1960).

20. Van Ness, H. C., Ind. Eng. Chem. Fund., 8, 3, 464 (1969).

21. Miller, G. W., "Adsorption Equilibria and Performance of a Pressure Swing Adsorption Air Separation Unit," MS Thesis, The Ohio State Univ., Columbus, Ohio (1984).

22. Sips, R. J., J. Chem. Phys., 16, 490 (1948).

23. Peterson, D., "Influence of Presorbed Water on the Sorption of Nitrogen by Zeolites at Ambient Temperatures" in Adsorption and Ion Exchange with Synthetic Zeolites, W. H. Flank (ed.), (1980).

24. Myers, A. L., "Adsorption of Pure Gases and their Mixtures on Heterogeneous Surfaces," in Fundamentals of Adsorption, A. L. Myers and G. Belfort (eds.), Engineering Foundation, p.365 (1984).

25. Breck, D. W., Zeolite Molecular Sieves, John Wiley and Sons, New York (1974).

PURIFICATION OF LIQUID ETHYLENE DICHLORIDE BY ADSORPTION

Ajay Sood and Hubert L. Fleming ■ Aluminum Company of America, Alcoa Laboratories, Alcoa Center, PA 15069

The separation of acidic contaminants from liquid ethylene dichloride has been examined. Through the use of a chemically modified alumina specifically designed for this application, it was found that packed bed adsorption could remove hydrochloric acid and ferric chloride to below one ppm in the effluent. Residual water is also significantly reduced.

Kinetics and thermodynamics of the adsorption process for each adsorbate are examined. Both dynamic columns and batch cells were evaluated in order to characterize the separation as a function of various process parameters. Some definition of process reversibility and potential regeneration schemes are given.

There is currently a trend in the petrochemical industry toward developing alternative processes to distillation. Because of the high energy costs for distillation, and the fact that it remains the predominant separations technique, process alternatives in this area are seen as a critical factor in remaining cost competitive. Separation of contaminants from high-boiling liquid monomers via alternative processes to distillation is seen as a significant area for improvement. Design of an adsorption process for the cleanup of ethylene dichloride, a large commodity petrochemical is a relevant example of this trend.

Using proprietary adsorption media and processing developed by Alcoa, a number of petroleum companies have commercialized units for EDC purification. Called Selexsorb-HCl, a material has been developed, characterized, and translated to plant production which exhibits superior performance for the removal of contaminants from ethylene dichloride.

Aluminum Company of America, Pittsburgh, Pennsylvania

Ethylene dichloride (EDC), also known as 1, 2-dichloroethane, is a commodity chemical used extensively in two industries. Because of favorable heat transfer properties, it has found acceptance as a refrigerant in cooling systems. A much greater application is as the feed material for production of polyvinylchloride (PVC). PVC is currently the third largest tonnage engineering plastic, behind polyethylene and polypropylene, and accounts for most of the world's use of EDC (1). In conventional petrochemical processing, EDC is converted to VCM, vinyl chloride monomer. The monomer, in turn, is polymerized to form the desired product, PVC, as shown in Figure 1.

During typical processing of the EDC, the liquid becomes somewhat degraded - possibly mechanically by pumping, thermally by heat generated in the flow system, and/or chemically by ultraviolet light. The 99.95% pure EDC degrades during processing to VCM by the petrochemical company. It is also possible that water is formed as a by product in the oxychlorination route to EDC, or left during the washing of the EDC in an attempt to remove the iron chloride. When the

EDC degrades, low levels of hydrochloric acid and water may be formed. Iron (as ferric chloride) is also found in most units since it is used as a liquid phase catalyst in the direct chlorination of ethylene to EDC. This is shown in Figure 1. It is believed that typical levels of these contaminants are 200 to 600 ppm H_2O, 50 to 100 ppm HCl, and one to five ppm Fe (predominately Fe^{3+}). In order to process the EDC, as well as to avoid downstream corrosion, the iron and HCI must be removed to less than one ppm. Removal of the water is questionable and is generally considered to be of secondary importance. It is agreed, however, that some reduction of the water loading is necessary for optimum performance.

EDC contamination phenomena were investigated so that the variables affecting contaminant levels were understood. Second, adsorption phenomena of the various contaminants were examined as function of alumina surface chemistry so that an appropriate adsorbent could be designed. Third, dynamic simulation of the EDC cleanup process was undertaken. Adsorber dynamics were defined for transfer of the process to potential end users.

Separation thermodynamics and equilibria were defined for Selexsorb-HCl, the optimum adsorbent in terms of multi-component isotherms. Transport phenomena, including pore kinetics were also examined in the same experimental system as a function of process variables. In the process simulations, dynamics of the purification process were characterized and compared with the static predictions. Following adsorption, desorption phenomena were studied, leading to various modes of regeneration. Autopsy analyses of the spent adsorbent columns were made in order to provide further information about the media. This paper will report only those results relevant to operation of a commerical process.

EXPERIMENTAL

Static Tests

Bench work for determining the adsorption isotherms, adsorbent/ capacities, adsorption kinetics, temperature effects, and regenerated adsorbent isotherms was performed in static (batch) tests. Batch systems are advantageous in that the impact of changing process parameters such as time, temperature, and adsorbate concentration could be rapidly assessed with minimal use of materials. Before performing the static test, an EDC solution containing the appropriate level of HCl, water, and ferric chloride contaminants was prepared. High purity ethylene dichloride was obtained from Fisher Scientific Company and intentionally doped with contaminants.

Static adsorption was conducted by contacting the adsorbent with EDC solution in a glass bottle with a Teflon lined cap. For the isotherm measurements, the mixture was well stirred for 24 hours, the liquid layer decanted and analyzed for HCl, Fe, and H_2O. From the initial and final concentrations, the equilibrium adsorption capacity (mg adsorbate/g adsorbent) was calculated. The bottles were placed in a constant temperature bath to obtain data at higher temperatures. For specific process definition, only two temperatures were investigated: room temperature (25°C) and 40°C. Adsorption capacities at 40°C were measured: (1) since ambient temperatures in some locations may approach this level, and (2) because temperature dependency of adsorption thermodynamics provides additional information regarding mechanisms of adsorption.

For the kinetic uptake measurements, the adsorbate/ adsorbent mixture was stirred for shorter times ranging from ten minutes to two hours and the adsorbate concentration changes measured as a function of time. Initial rates of adsorption were calculated as a function of adsorbent sizes and pore distributions. The impact of pore diffusion on adsorption kinetics was determined for two sizes of adsorbent: one eighth inch spheres, the adsorbent size defined for commerical application, and -48/+100 Tyler mesh, a size found to be small enough to represent negligible pore resistances due to the nominal transport length.

Process Dynamics

Dynamic simulation of the EDC purification process was demonstrated in a specially designed Process Development Unit (PDU). A schematic flow sheet of the PDU is shown in Figure 2. The unit comprised four adsorption colums in parallel, each with its own feed tank, feed pump, rotameter, and level controller. All four feed tanks were supplied with EDC solution from a common mixing tank MU1 having a 40 gallon capacity. All effluent from the four columns was collected in two 20-gallon hold tanks, HT1 and HT2.

The EDC solution was extremely corrosive due to the levels of HCl and $FeCl_3$ added to it in the mixing tank. Therefore, glass was used for all columns, tanks and level controllers. Teflon tubing was used in all locations where it would contact EDC solution. The pumps were laboratory size fluid metering pumps. A special feature of these pumps was that the EDC solution came in contact with a chemically inert ceramic α-alumina piston and cylinder head assembly only. The piston seals were precisely-cut fluorocarbon sheets to resist wear, abrasion, heat, and chemical attack. All valves and fittings were made of SS 316; some slight corrosion of the SS fittings was observed after 45 days of continuous operation. However, it is believed that additional iron introduced in the EDC solution due to corrosion of the SS fittings was minuscule.

Fresh EDC was pumped from its container to the mixing tank by a pump. A stock solution of the contaminants was prepared separately from concentrated hydrochloric acid (∼38% HCl) and anhydrous ferric chloride. No additional water was added to the feed. The amount of stock solution necessary to spike the fresh EDC to the desired feed/EDC composition was weighed and added to the mixing tank. The EDC solution was thoroughly mixed for 20 minutes to ensure that the contaminants were uniformly dispersed in the feed solution. The mixing was done with a Teflon paddle-type stirrer driven with an air motor.

Once the adsorption process was started, column effluents were collected in the hold tanks and recycled to the mixing tank to make up more feed. The recycled EDC solution was analyzed for HCl and Fe before spiking it with the stock solution. Negligible HCl and Fe were observed in the column effluents for all runs. Thus, the amount of stock solution used for spiking was held constant.

EDC solution was pumped from feed tank to a column, where it passed through a packed bed of adsorbent in a downflow direction. Each feed metering pump had a micrometer so that the flow could be precisely set within a range of zero to 450 cc/m. The flow rate was carefully monitored with the rotameter to ensure that it stayed at the desired level. A manometer-type arrangement was used to maintain approximately two inches of liquid head above the adsorbent. The incoming EDC solution was introduced in through Teflon tubing and released near the top of the adsorbent bed; this minimized splashing of EDC at top of the column. All columns were two inches in diameter and six feet high. Column effluent was sampled through a globe valve at the column exit.

Dynamic column test conditions used during the adsorption tests are shown in Table 1. A range of contact times, space velocities, stream compositions, and temperatures could be evaluated in these four parallel units. Table 1 lists representative conditions for PDU, as well as simulated commercial operations.

A typical column was charged with 704 g of Selexsorb-HCl (oven dried at 110°C), giving a packed bed volume of 1155 cc and a bed height of 57 cm; the volume occupied by the solids in the bed, i.e., the volume which excludes the interparticle and intraparticle porosity, was 920 cc. The space velocity of two hr^{-1} was based on this solids bed volume of 920 cc. The calculated packed density was 0.609 g/cc. The EDC flow rate to the column was held at 30.6 cc/m in order to get a 30-minute contact time.

Because of the safety aspects in dealing with ethylene dichloride (toxic, flammable, hazardous, potentially carcinogenic, and difficult to detect), the PDU was placed in a specially designed laboratory inside a large containment hood. This hood was equipped with specific detectors for potential effluents, as well as having the capacity to deal with spill and/or leaks of large volumes of contaminated EDC. All personnel in contact with the apparatus wore particular safety equipment, including self-contained breathing apparatus.

Analytical Techniques

Three key analyses were performed on most EDC samples. These were the measurement of total iron, chloride ion, and water concentrations. Iron concentrations were determined by measuring absorption at 366 nm with a Bausch & Lomb Spectrometer 70. The instrument was calibrated with EDC samples having a known amount of iron and the calibration curve was linear over the zero to six ppm range. Minimum detection limits of the method were estimated as < 0.1 ppm Fe.

Chloride ion concentration in EDC samples was determined by extracting the ions into an aqueous medium and measuring the chloride concentration in the aqueous solution by ion chromatography. EDC is immiscible in the water layer. The method is applicable for chloride ion concentrations from 0.1 to 120 mg/L in the EDC, with a minimum detection limit of < 0.2 ppm HCl. Higher concentrations could be analyzed by adjusting the aliquot size of the EDC sample.

Water analyses in the EDC samples were done by the Karl Fischer method. Using K-F solution of iodine, sulfur dioxide, pyridene, and methanol, iodine is reduced to hydrochloric acid. Acid amount is dependent on the amount of water present. Both manual and automated (Aduatest) K-F titrators were tried. Aquatest is a coulometer designed specifically for K-F titrations of water. Less variability was found with the Aquatest unit, due to automatic control of reagent addition, and hence it was used exclusively in later experimentation.

RESULTS AND DISCUSSION

Adsorption Isotherms

Comparison of the static adsorption capacities of Selexsorb-HCl is shown in Table 2 at two temperatures of 25°C and 40°C. All isotherms were measured as multi-component, rather than singly to be more relevant. The capacities for Fe, HCl, and water were compared at nominal EDC solution concentrations of five ppm Fe, 80 ppm HCl, and 300 ppm H_2O because of industrial interest in these conditions. Typically, HCl is present in excess of Fe, with H_2O being present in widely variable concentrations.

At 25°, the Selexsorb-HCl adsorbent had a capacity of 0.8% Fe, 34% HCl, and 8% H_2O by weight. Individual isotherms are shown in Figures 3 to 5. Different adsorption models were used to fit each of the three contaminants. In all cases (published here, as well as unpublished), iron adsorption follows the Langmuir model, HCl adsorption follows the Redlich-Peterson model, while water adsorption fits the Freundlich model.

Iron adsorption was found to be strong and specific (fitting Langmuir thermodynamics), while hydrochloric acid displayed complex behavior, depending upon solution concentration. At low concentrations of HCl conventional chemisorption occurred. At the higher levels, the second mechanism of liquid-solid reaction to form aluminum chloride complex was evident, resulting in modified Langmuir behavior toward Freundlich (i.e., Redlich-Peterson) fits. Water adsorption was quite complex, with considerable hydrogen bonding and formation of multilayers, thus a Freundlich fit of the data.

An increase in adsorption temperature to 40°C was found to dramatically increase iron and HCl capacity and some increase in H_2O capacity. As seen in Figure 3, Fe capacity increases, without an asymptotic capacity being determined. Unfortunately, solutions with higher Fe concentrations than 3.3 ppm were difficult to prepare at the elevated temperature. The same phenomenon is observed (Figure 4) with HCl, with capacities increasing by more than 100% at the slightly elevated temperature. Once again, it was difficult to obtain equilibrium data at higher HCl levels due to solution instability at 40°C.

Water uptake was similar to both Fe and NCl in its temperature dependence. With capacities more than doubling, the region in concentration where a second regime of removal appears to begin moved to lower H_2O solution concentrations. That is, at higher temperatures, less water is required to obtain high adsorption capacities, consistent with the suggestion of a second, solid-liquid reaction mechanism being present.

<u>Adsorption Kinetics</u>

Rates of uptake for each adsorbate were measured statically in order to better characterize adsorption dynamics. Adsorbate/adsorbent mixtures were stirred for times ranging from 10 minutes to 24 hours and adsorbate concentration changes measured as a function of time (Figure 6). Initial rates of adsorption were calculated as a function of such parameters as adsorbent pore distribution, media particle size, and adsorber temperature. The impact of pore diffusion on adsorption kinetics was assessed using two sizes of adsorbent: one eighth inche spheres and 48 X 100 Tyler mesh material. Using 0.1 g of adsorbate stirred with 100 mL of EDC solution, results for Fe, HCl, and H_2O adsorption as a function of particle size at 25°C are shown in Figures 6 to 8, respectively.

Initial rates of adsorption measured with the 48 X 100 mesh material were generally much higher than those with the one eighth spheres. Significant pore diffusion limitations were found, retarding the rate of mass transfer in the larger size spheres. The diffusional resistances are important at short contact times such as the 0.5 h contact time used in the dynamic column experiments (now being used commercially). At sufficiently long contact times, e.g. 24 hours, the adsorbate concentration approaches thermodynamic equilibrium and is approximately the same for both sizes.

A good approximation of the mass transport effectiveness factor, η, was calculated by ratio of the initial rates using the two sizes. This procedure was found to be valid for distribution of the diffusional resistance due to pore structure by using smaller particles (< 325 mesh), where no further retardation of transport was observed (Table 3). For one eighth inch Selexsorb-HCl, the effectiveness factors for Fe, HCl, and H_2O were 0.14, 0.57, and 0.68 respectively. Although pore diffusion is clearly the predominant resistance to mass transport in the adsorption process, mass transfer is more than acceptable for a process having a relatively viscous liquid

liquid as the continuous phase.

Dynamic Column Performance

Effluent profiles for the PDU packed with one eighth inch spheres of Selexorb-HCl at 25°C are shown in Figures 9 and 10. For the runs shown here, the composition of the feed EDC stream averaged 5.69 ± 0.60 ppm Fe, 77 ± 16 ppm HCl, and 867 ± 5 ppm H_2O. Virtually complete Fe and HCl removal was obtained in the column throughout 44 days of continuous operation. Column effluent averaged roughly < 0.1 ppm Fe, < 0.4 ppm HCl, while H_2O began to strip through, increasing from 300 to 600 ppm. Figure 10 is a scale expansion of the zero to one ppm region on effluent HCL and Fe concentrations. No Fe or Hcl breakthrough was observed in the column during the entire operation. Because of time constraints, these columns were typically not run to complete breakthrough. Although not presented here, autopsy analyses of column media throughout the bed, excluding the top layer, following the runs showed significant capacity remaining in the beds for HCl and Fe at this point. As seen in Figure 10, effluent HCl and Fe levels remain constant over time with constant breakthrough, while H_2O streams through early, with a reasonably constant H_2O level in the effluent.

Removal efficiencies for the three contaminants were calculated based on the mean feed and effluent concentrations. The average removal efficiency was 98.2% for Fe, 99.4% for HCl, and 17.9% for H_2O. The results indicate virtually complete Fe and HCl removal, since the Fe and HCl remaining in the effluent were close to the detection limits of the analytical techniques used.

Analyses of the spent adsorbent media generally confirmed static data in terms of equilibrium capacities. Hydrochloric acid loadings of 26 to 30 wt% were found on the solids, as well as 0.8 to 3.2 wt% Fe. Bulk water content on the solids is difficult to measure because of the additional presence of surface hydroxyls.

Cumulative pore distributions for Selexsorb-HCl as a function of contact time on stream are given in Figure 11. Completely spent material had 0.22 cc/g total pore volume remaining, most occurring bimodally at 50 Å and 10,000 Å. For reference, the fresh material exhibits porosity over the entire pore range with the largest peaks at 40 Å and 15,000 Å, although there is quite a bit up to 300 Å. Random samples of partially spent material exhibited minimal loss ob total pore volumes, only a reduction from 0.65 cc/g to 0.58 cc/g. Microporosity is shifted to slightly larger size, indicative of a chemical dissolution process, with the subsequent growth of macroporosity in the 15,000 to 20,000 Å range.

Regeneration Schemes

Three viable processes were developed for treatment of the spent adsorbent bed. For plants where neither chemical nor thermal regeneration is desired, disposal of the spent bed can be relatively economical because of the exceptionally long life of the adsorbent, ranging commercially from one to three years. An apparently optimal procedure for bed disposal is the contacting of one bed volume of one percent NaOH with the bed to neutralize excess hydrochloric acid, drainage, then discharging the bed out of the unit. It was found that more concentrated caustic results in some bed dissolution, with a subsequent loss of flowability. More dilute NaOH can leave residual acid which also begins degradation and fusing of the bed.

Chemical regeneration can be successfully accomplished via a methanol (or any other alcohol) rinse following bed drainage. Using one to three bed volumes of methanol, followed by an ambient air flush, the bed can be continuously reused. As seen in Table IV, the relatively aprotic nature of the alcohol minimizes bed dissolution, while resulting in 50% to 75% removal of HCl, Fe, and H_2O.

Adsorption capacities for regenerated beds are given in Figures 12 and 13 for HCl and Fe, respectively. It is seen that the materials retain their capacities, particularly in the critical region of low adsorbate (0.5 to 5.0 ppm for Fe) concentrations. In fact, the capacity for iron appears to increase slightly after regeneration.

Thermal regeneration is also practical. As shown in Figure 14, virtually all adsorbed species (except iron) are removed at less than 300 °C. In fact, using a mass spectrometer coupled to the desorption apparatus, it is seen that fully 60% of the HCl is removed at the lower temperature (< 200°C), followed by the bulk of the water. Desorption rates peak at over 10%/min., which is quite good. No iron is removed. However, adsorption for each adsorbed species is regained (Figures 15 to 16). There is some loss of affinity for iron at the lower (< 3 ppm) solution concentrations. This behavior was repeated for a few multiple cycles in batch studies, but not in columns, due to limitations in time and volume of feed liquid necessary.

Conclusions

Through the use of an alumina chemically modified specifically for hydrochloric acid and iron retention, it is possible to economically purify ethylene dichloride via isothermal adsorption. Process dynamics, with multiple options for regeneration, have been successfully demonstrated. Although designed for EDC purification, the media and process should be quite applicable to any organic liquid containing HCl, H_2O, and/or Fe as adsorbates.

LITERATURE CITED

1. Kirk-Othmer Encyclopedia of Chemical Technology, Vol.23, John Wiley & Sons, New York (1983).

2. S. Joshi, J. L. Humphrey, J. R. Fair, "Adsorptive Drying of Organic Liquid--an Update," Proceedings Petroenergy 1986, Houston (May 1986).

TABLE 1
DYNAMIC COLUMN TEST CONDITIONS

L.H.S.V. (hr^{-1})	2
Adsorbent Wt. (grams)	704
Solids Bed Vol. (cc)	920
EDC Flowrate (cc/m)	30.6
Bed Temperature, °C	25
Adsorbent Packed Bulk Density (g/cc)	0.609

TABLE 2
ADSORPTION CAPACITIES OF SELEXSORB-HCl

Specie	Equilibrium Capacity, wt% 25°C	40°C	Adsorption Model
Fe	0.8	4.5	Langmuir
HCl	34	32	R. Peterson
H_2O	8	16	Freundlich

TABLE 3
KINETICS RESULTS
(Well Stirred Batch, 25°C)

SELEXSORB-HCl rAD (ppm/hr)	Fe	HCl	H_2O
1/8"	5.9	276	294
48 x 100 mesh	42	486	435
Effectiveness Factor	0.14	0.57	0.68

TABLE 4
METHANOL REGENERATION OF SELEXSORB-HCl

Adsorbate	Solution Concentration (mg/L)	Solid Concentration (wt%)	Regenerated Solid Concentration (wt%)	Removal Efficiency (%)
Iron	459	1.00	0.44	56.0
Chloride	10,700	32.1	8.18	74.5
Water	8,270	18.2	9.1	50.0

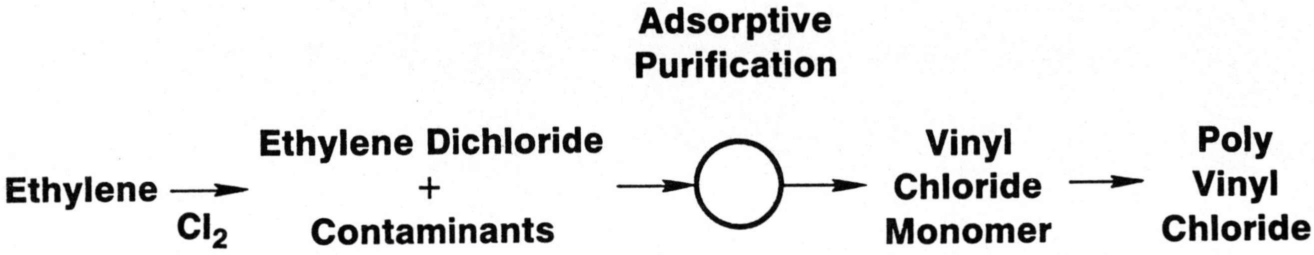

Figure 1. PVC production process.

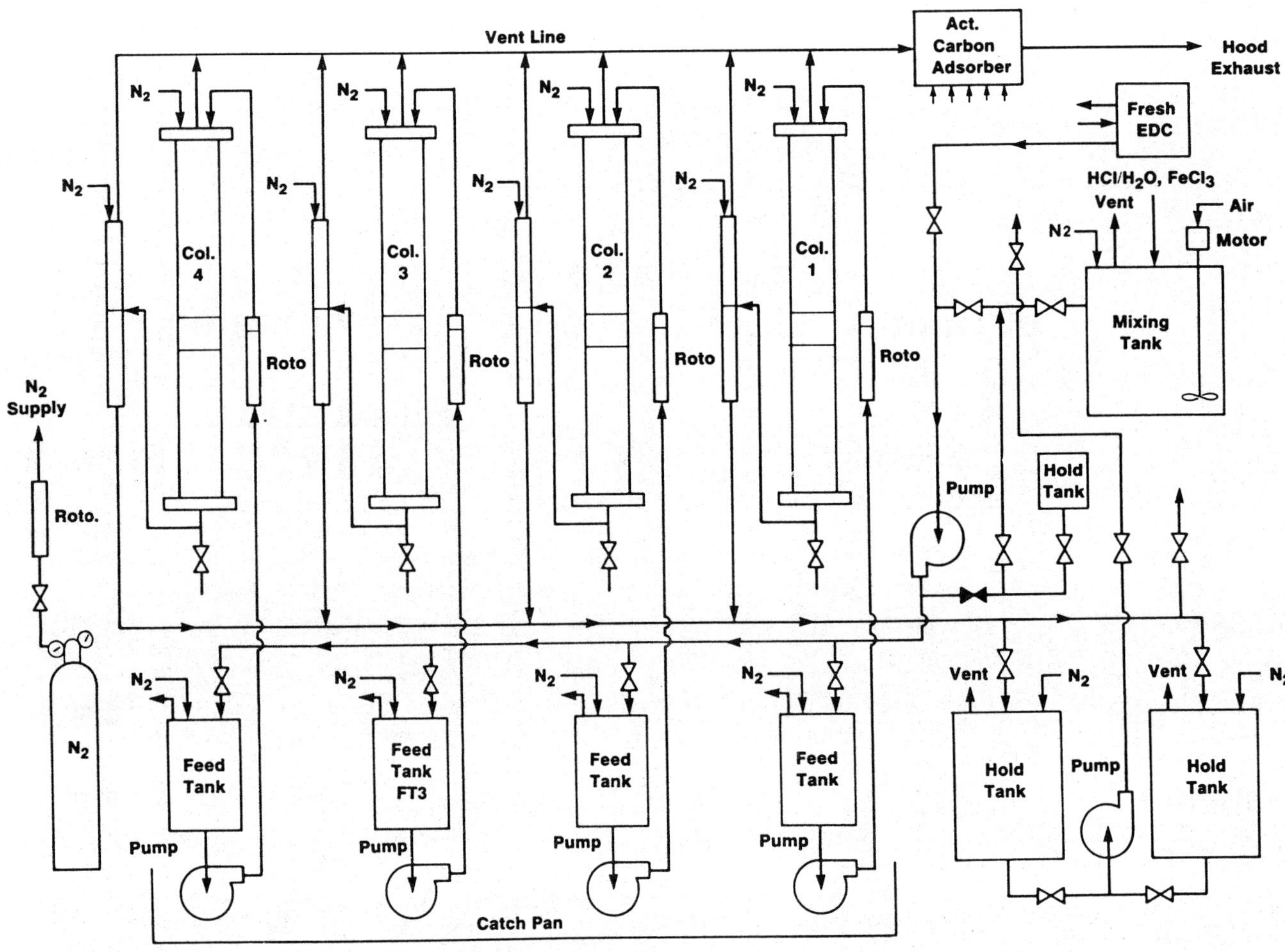

Figure 2. Schematic flowsheet of EDC adsorption unit.

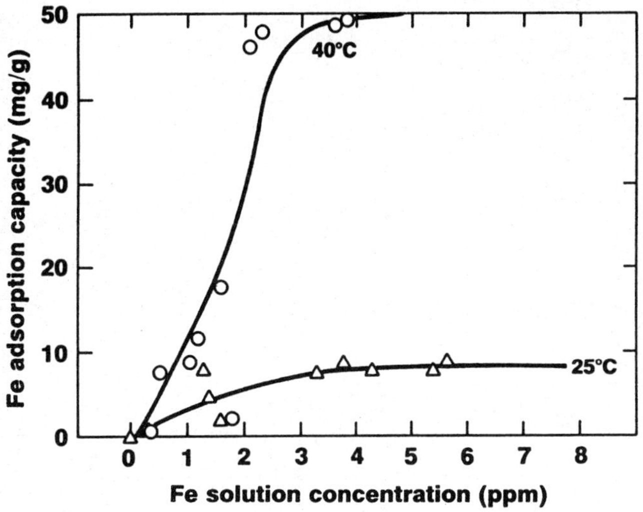

Figure 3. Adsorption isotherm - Fe.

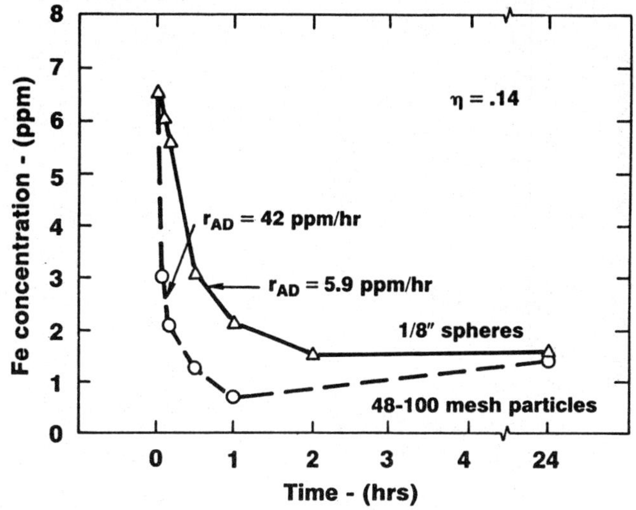

Figure 6. Adsorption kinetics - Fe.

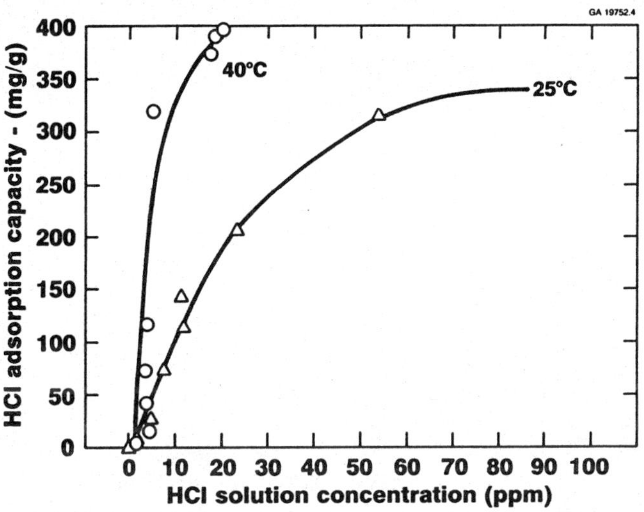

Figure 4. Adsorption isotherm - HCl.

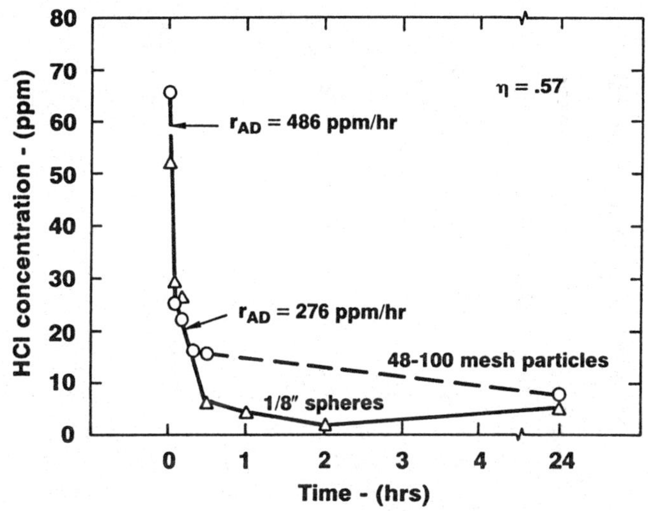

Figure 7. Adsorption kinetics - HCl 0.1g/100 ml - 25°C.

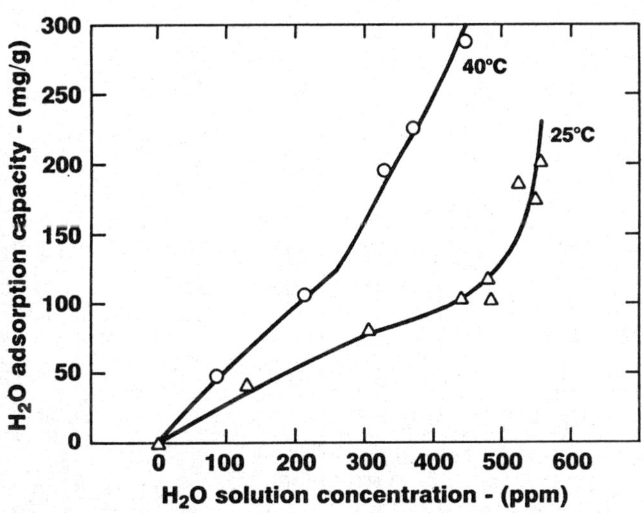

Figure 5. Adsorption isotherm - H_2O.

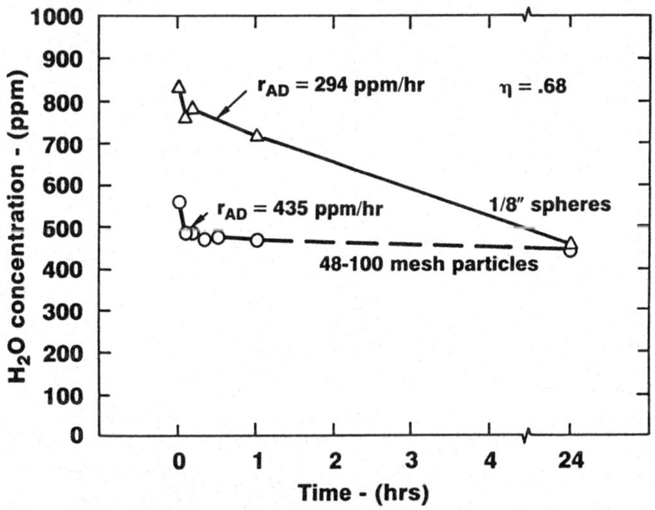

Figure 8. Adsorption kinetics - H_2O 0.1g/100 ml - 25°C.

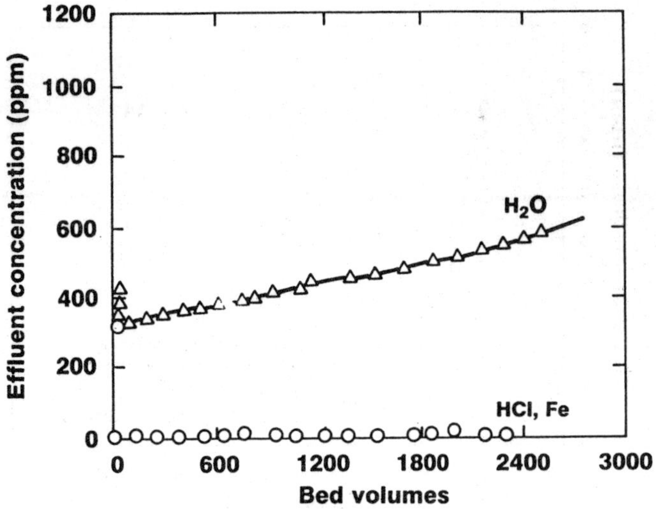

Figure 9. Dynamic column effluents.

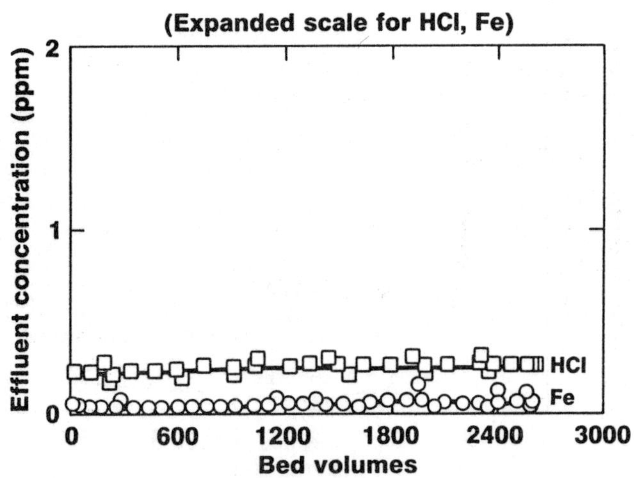

Figure 10. Dynamic column effluent.

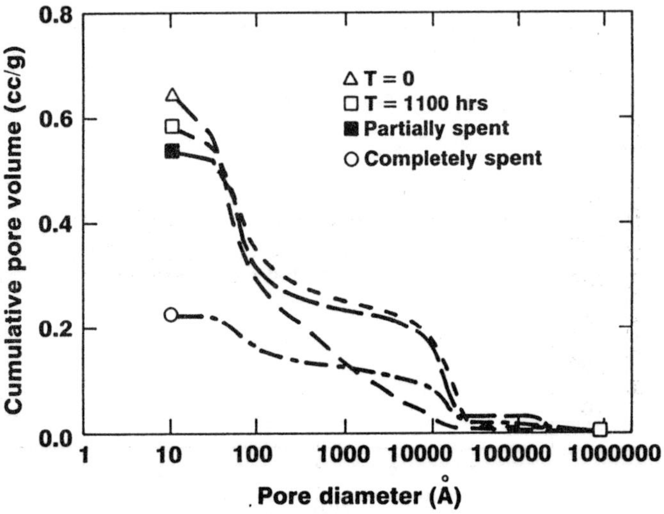

Figure 11. Cumulative pore distributions.

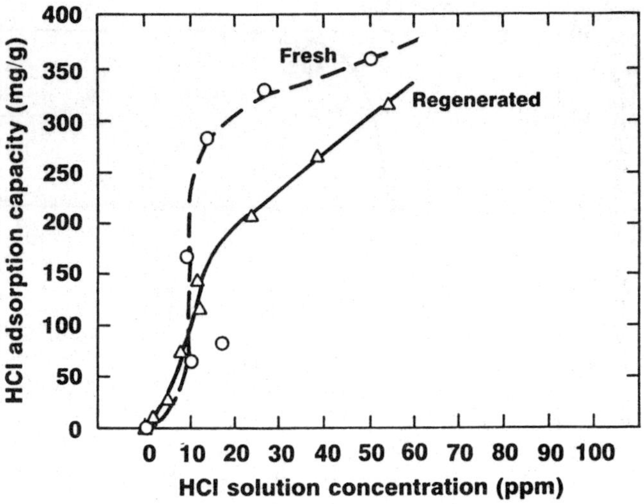

Figure 12. Adsorption isotherm - HCl Fresh vs methanol regenerated Selexsorb-HCl.

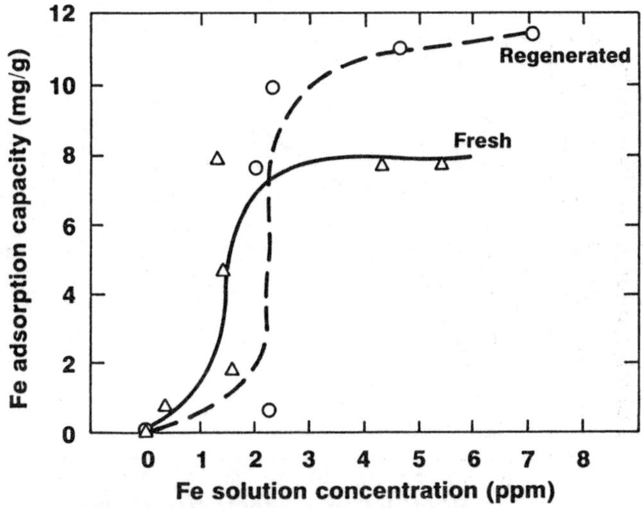

Figure 13. Adsorption isotherm - Fe Fresh vs methanol regenerated Selexsorb - HCl.

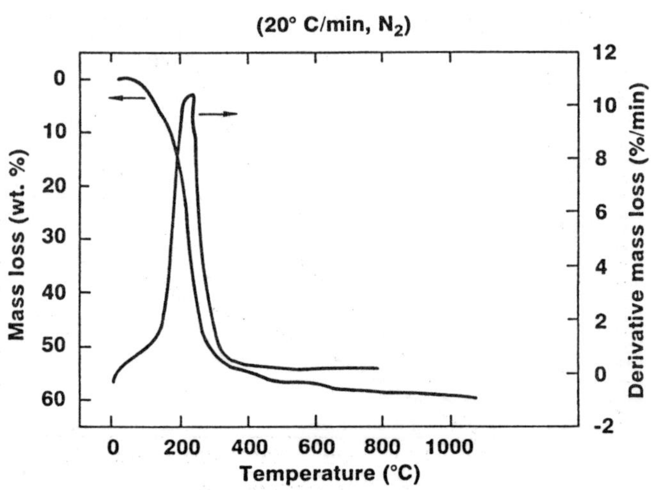

Figure 14. Thermal desorption of Selexsorb - HCl.

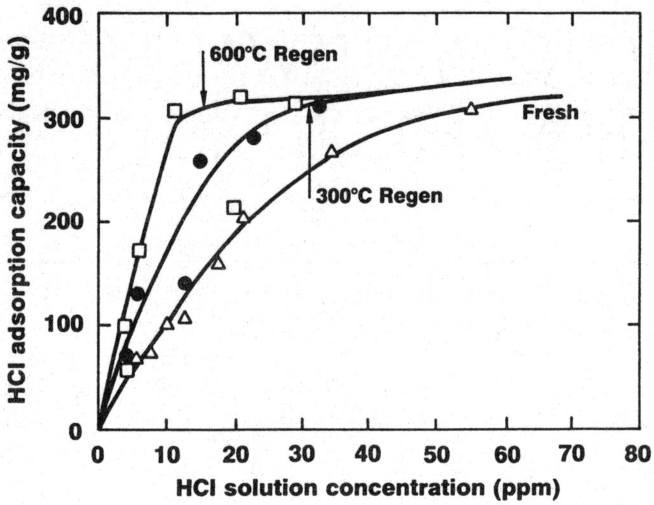

Figure 15. Adsorption isotherm for HCl as a function of thermal regeneration.

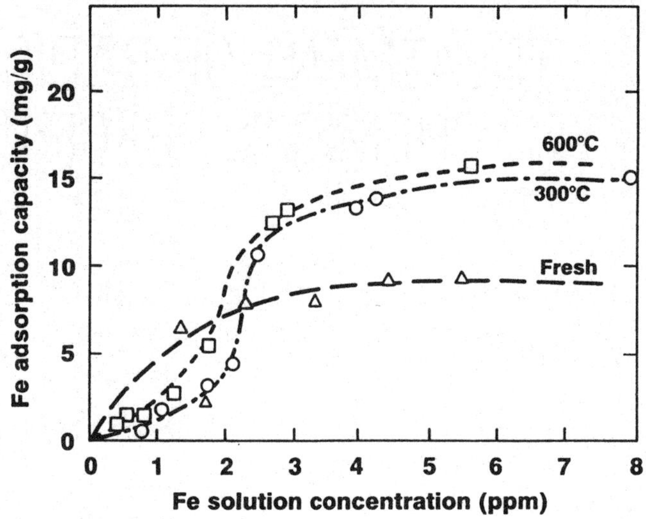

Figure 16. Adsorption isotherm for Fe as a function of thermal regeneration.

TREATMENT OF RADIOACTIVE WASTEWATERS BY CHEMICAL PRECIPITATION AND ION EXCHANGE

Sharon M. Robinson, John M. Begovich, Clifton H. Brown, Jr., David O. Campbell and Emory D. Collins ■
Chemical Technology Division, Oak Ridge National Laboratory*, Oak Ridge, TN 37831-6316

Precipitation and ion exchange methods are being developed at Oak Ridge National Laboratory to decontaminate wastewaters containing small amounts of ^{90}Sr and ^{137}Cs while minimizing waste generation. Using actual wastewaters, distribution coefficients have been determined for strontium and cesium, as functions of Ca, Na, and Mg concentrations, from bench- and pilot-scale data for ion exchange resins and zeolites. Models have been used to estimate the total amount of waste that would be generated at full-scale operation. Based on these data, four process flowsheets are being tested at full-scale.

INTRODUCTION

Improved chemical precipitation and/or ion exchange methods are being developed to decontaminate process wastewater at Oak Ridge National Laboratory (ORNL) while concentrating the radioactive materials into a nonhazardous solid waste form that can be safely stored for permanent disposal. The process wastewater, which is slightly contaminated with ^{90}Sr and ^{137}Cs, has been routinely processed at ORNL at a rate of 6 to 10 L/s for 25 years by chemical precipitation and/or ion exchange methods. Although these treatments have decontaminated the process wastewater sufficiently for release to the environment, technical and regulatory problems associated with the formerly used low-level radioactive waste (LLW) disposal method have prompted ORNL to seek improved decontamination methods that will reduce the volume of LLW generated.

A wide variety of potential chemical precipitation techniques and ion exchange materials were considered for possible use in laboratory-scale screening tests. Initial scouting tests resulted in the selection of two caustic—soda-ash softening processes to be tested in conjunction with ion exchange materials in proposed process flowsheets. Each softening process reduces the calcium and magnesium ion concentrations (the major ions that compete with the radionuclides for adsorption sites on ion exchange materials) in the process wastewater from ~50 mg/L Ca and 10 mg/L Mg to <5 mg/L of each. One process has the advantage of minimizing the sodium concentration in the softened water; the major advantage of the second process is its ability to accommodate fluctuations in the feed stream more easily.

Experimental, small-scale column tests that contained 6.5 to 20 mL of ion exchange material were made to determine distribution and mass transfer coefficients of 16 commercially available zeolites and organic cation exchange resins as a function of the Ca, Mg, and Na concentrations in the feed stream. Results from these tests indicated that for the case of unsoftened process water, most of the zeolites have high ^{137}Cs distribution coefficients (up to 23,000) but the sodium concentration is in the range of 20 to 40 mg/L. However, the zeolites were found to have lower distribution coefficients for strontium, and the sorption kinetics are relatively slow. The chabazite (a type of zeolite) samples had distribution coefficients of 2000 to 3000 for ^{90}Sr when column residence times of 10 min were used. Several materials are good sorbents for strontium when the feed has been pretreated by either of the softening processes (i.e., clinoptilolite zeolites and weak-acid resins have ^{90}Sr distribution coefficients of 10,000 to 25,000).

Calculations were made to model the performance of the more-promising sorbents

* Operated by Martin Marietta Energy Systems, Inc., for the U.S. Department of Energy under contract DE-AC05-84OR21400.

and to predict the results for full-scale operating conditions. Based on these results, four process flowsheets have been proposed for pilot- and full-scale testing. A plant-scale water-softening unit has been installed in the existing treatment plant upstream of columns containing a strong-acid cation exchange resin, HCR-S, produced by Dow Chemical Company. Lowering the total hardness of the feed from 150 mg/L (as $CaCO_3$) to <10 mg/L has extended the life of the columns from 400 to 12,000 bed volumes (bv). Treatment of unsoftened process water with a synthetic chabazite, Ionsiv IE-95 (manufactured by Union Carbide), has also been tested at full-scale. Two 3.7-m^3 columns, operated in series with a feed rate of 7.6 L/s, confirmed laboratory-scale data. Tests conducted on 10% plant-scale equipment, which consists of four columns in series containing a cheaper natural chabazite, PDZ-300 (supplied by Tenneco Specialty Minerals), have performed better than those conducted with the IE-95. Tests are planned to evaluate the other promising sorbents and to develop treatment and packaging processes of the spent sorbents for permanent disposal.

COMPOSITION OF ORNL PROCESS WASTEWATER

The Process Wastewater System at ORNL is used to collect water that has the potential to be contaminated. For example, water used in heat exchangers for radioactive liquids could become contaminated if leakage occurred within the exchangers, and it is, therefore, discharged via the Process Wastewater System. The process wastewater is collected in an equalization basin for subsequent treatment prior to discharge to the environment. The process wastewater contains a number of trace radionuclides, shown in Table 1, and relatively large amounts of competing ions (representative of city water and local groundwater in Oak Ridge, Tennessee), as shown in Table 2. The major chemical constituents are calcium, sodium, and magnesium bicarbonates, and the major radionuclides are ^{90}Sr and ^{137}Cs. The ^{90}Sr is the more hazardous contaminant because of its potential for introduction into the human and animal food chain. It tends to be the limiting ion in most processing alternatives.

Table 1. Radiochemical Composition of Process Wastewater

Radionuclide	Concentration Bq/L[a]
Gross alpha	5
Gross beta	6000
^{60}Co	25
^{90}Sr	4000
$^{95}ZrNb$	50
^{106}Ru	10
^{137}Cs	400

[a]1 Bq = 1 disintegration per second = 2.7 x 10^{-11} Curies.

Concentrated spikes of radioactive materials have occasionally entered the feed stream due to the decontamination of research facilities, leakage from equipment, etc. Variations in feed composition between 2000 and 8000 Bq/L for ^{90}Sr and 300 and 1000 Bq/L for ^{137}Cs have not been uncommon. The effects of such concentration changes on treatment systems are discussed below.

POTENTIAL TREATMENT PROCESSES

Several processing methods have been used to decontaminate the process wastewater at ORNL. Prior to 1975, it was treated by means of a conventional lime-soda precipitation—clarification method (Browder, 1959). The water was not sufficiently decontaminated when the limits for release to the environment were lowered, and an improved scavenging-precipitation ion exchange (SPIX) process was initiated (Chilton and Lasher, 1982). The new process included a sodium hydroxide precipitation step performed with ferrous sulfate, followed by clarification, polishing filtration, and ion exchange treatment. The ion exchange columns contained a bifunctional phenolic-carboxylate resin, Duolite CS-100, in the sodium form. The resin was regenerated with 0.5 $\underline{M}$ HNO_3 after treating ~2000 bv of process wastewater. The eluate was concentrated by evaporation and transferred to the LLW Treatment System at ORNL for permanent disposal. The sludge from the clarifier was stored in a rubber-lined pond. This process was discontinued, however, because of resin degradation and sludge disposal problems.

From 1981 to 1985, the processing method was a filtration—ion exchange (FIX) process. In this process, a more durable strong-acid resin, HCR-S, in the hydrogen form (prepared by Dow Chemical Company), was used without the clarification step. The process generated more waste (regeneration was required after ~400 bv), but it was easily operated and accommodated fluctuations in the feed stream flow rate and composition. The larger volume of LLW was handled without problems until new disposal methods for LLW were implemented.

Table 2. Chemical Composition of Process Wastewater

Cation	Conc. mg/L	Anion	Conc. mg/L	Parameter	Conc. mg/L
Ca	40	HCO_3	93	pH	8.8
Mg	8	SO_4	23	Total Dissolved Solids	250
Na	5	Cl	10		
K	2	NO_3	11	Total Suspended Solids	3
Si	2	CO_3	7		
Sr	0.1	F	1	Total Hardness[a]	133
Al	0.1			Alkalinity[a]	125
Fe	0.1			Chemical Oxygen Demand	6
Zn	0.1				
				Total Organic Carbon	12

[a]As $CaCO_3$.

Improved methods were investigated for treatment of process wastewater based on these experiences, the development of an ion exchange process to decontaminate high-activity-level water at Three Mile Island (Collins et al., 1986), and other published data. Mercer (1967) has published equilibrium constants for Cs^+ and Sr^{2+} and ion exchangers as a function of several competing ions, including H^+, NH_4^+, Na^+, K^+, Mg^{2+}, and Ca^{2+}. His data indicate that sorption of ^{90}Sr and ^{137}Cs ions from ORNL process water would be inhibited by calcium ions, unless the water is first softened by a precipitation/clarification step. In the latter case, the major inhibiting ion would be sodium. Based on previous experience, a zeolite was expected to remove ^{137}Cs without difficulty. However, finding an ion exchange resin capable of separating ^{90}Sr from calcium ions was considered doubtful. Thus, a general flowsheet was developed for testing which included a zeolite to remove ^{137}Cs and a softening process to remove calcium, followed by the use of an ion exchange resin to remove ^{90}Sr.

MODEL

The experimental breakthrough data were fitted by means of the mathematical J-function, using a constant separation factor model developed by Thomas (Hiester et al., 1973; Helfferich, 1962) in order to predict distribution coefficients and the performance of columns under different operating conditions (i.e., longer columns, columns in series, and different flow rates). The general Thomas equation for the reaction kinetics of fixed-bed ion exchange is as follows:

$$-\left(\frac{\partial X}{\partial N}\right)_{NT} = \left(\frac{\partial Y}{\partial NT}\right)_N = X(1-Y) - RY(1-X), \quad (1)$$

where X and Y are the dimensionless concentrations of the solute ion in the fluid and solid phases, respectively, and R is the separation factor. The variable X is defined as C/C_o, where C and C_o are the concentrations of the solute ion of interest in the effluent and feed, respectively. The variable Y is defined as q/q^*, where q is the actual concentration in the solid phase, and q^* is the concentration in the solid phase in equilibrium with fluid at the inlet concentration, C_o. When the concentration of the solute ion is small relative to the concentration of the replaceable ion in the feed, R approaches unity and the isotherm is linear.

The variable N represents the length of the exchange column in transfer units and is defined by the expression

$$N = K_d' \rho_b K_a/(f/v) \ , \quad (2)$$

in which K_d' is the distribution coefficient when $X = 1$, ρ_b is the bulk density of the ion exchanger, K_a is the mass-transfer coefficient characteristic of the system, f is the rate of flow of solution through the column, and v denotes the overall volume of the sorbent bed, including the void spaces. The throughput parameter, T, is defined to be:

$$T = (V/v)/K_d' \rho_b \ , \quad (3)$$

where V is the volume of solution processed through the column and V/v is the number of

bed volumes of solution which have passed through the bed.

When ρ_b is constant, the volume-based distribution coefficient is defined as $K_d = q_v/C_o$, where q_v is the concentration of the solute ion per unit volume of sorbent bed (sorbent plus void space) and C_o is the concentration in the feed. Equations (2) and (3) can then be expressed as

$$N = K_d K_a/(f/v) \quad , \tag{4}$$

and

$$T = (V/v)/K_d \quad . \tag{5}$$

When Equation (1) is integrated for ion exchange beds, assuming reversible second-order-reaction kinetics, the solution is:

$$X = C/C_o = \frac{J(RN,NT)}{J(RN,NT) + [1-J(N,RNT)] \exp[(R-1)N(T-1)]} \quad , \tag{6}$$

where J is a mathematical function (Hiester and Vermeulen, 1952) related to the Bessel function.

For the large values of RN, $C/C_o \doteq 0.5$ when $T = 1$ and is independent of RN. Therefore, K_d is approximately equal to V/v at the point where $C/C_o = 0.5$. This characteristic implies that plots of experimental data on logarithmic-probability graphs will be approximately linear. Therefore, K_d is approximately equal to V/v at the point where $C/C_o = 0.5$. This characteristic implies that plots of experimental data on logarithmic-probability graphs will be approximately linear. Therefore, K_d can be approximated by obtaining the 50% point on experimental breakthrough curves or by extrapolating experimental data on logarithmic-probability plots of C/C_o vs V/v. Values of R and N can then be obtained from the experimental data by iterating around Equations (5) and (6) and subsequently using these values to estimate performance at increased residence times and for columns in series.

LABORATORY SCREENING STUDIES

Several potential alternatives are available for use in each step of the general flowsheet described previously. Laboratory-scale scoping tests were conducted for each separate step of the proposed flowsheet to determine which options should be considered for further development. Sixteen inorganic and organic ion exchange materials were evaluated using untreated process wastewater as well as water softened by four processes.

Softening Tests

Small-scale batch tests were used to evaluate the softening processes used by the water-treatment industry for application in process wastewater treatment. Although the treatment of similar waste is standard in municipal applications, these industrial processes were developed without a consideration for the additional requirements created by nuclear materials, such as the increased cost of waste disposal and the need to minimize operator exposure to the waste. The traditional lime—soda-ash and caustic—soda-ash processes were evaluated, along with the less frequently used caustic precipitation process (Hamer et al., 1961). The scavenging-precipitation softening method used in the SPIX process was also evaluated. The latter is a caustic precipitation process in which ferrous sulfate is added as a flocculating agent and scavenger to help remove insoluble materials.

The softening processes were evaluated using 800-mL samples, and the waste was reacted for 75 min at 30 to 40 rpm via a paddle-type agitator. Such tests are standardly used for quality control of in-plant water-softening processes to determine parameters such as dosage requirements, pH, alkalinity, and floc time, but they cannot be used to predict actual operation in a full-scale plant (Foust and Aly, 1983). In actuality, full-scale operations usually soften water better than do laboratory-scale tests.

Each type of softening process was evaluated by varying the chemical composition and final pH between 10.5 and 12. Alum and ferrous sulfate were tested as potential coagulating agents, along with six synthetic, water-soluble polyelectrolytes with high to very high molecular weights and a range of anionic and cationic charge densities (Betz 1100, manufactured by Betz Laboratories, Inc.; Purifloc A-23, manufactured by Dow Chemical Company; and Percol 720, 726, and 728 and 757, produced by Allied Colloids). The reacting solutions were spiked with $CaCO_3$ to nucleate precipitation in an attempt to simulate a sludge blanket typically maintained in many continuous-operation precipitators. The sludge was dewatered in a laboratory-scale, single-frame filter press at 620 kPa (75 psig) for 1.5 h using Feon 162 Dynel filter cloth. The filter cake was air-dried for 30 min before removal from the press.

The optimum recipe for each process, based on softening ability and sludge characteristics, is listed in Table 3. A calcium hardness (measured in mg/L as $CaCO_3$) of less than 10 mg/L was considered acceptable, because lower values can be expected in full-scale operation. The polymers and alum produced granular flocs, but ferrous sulfate produced a fluffy, voluminous precipitate that was hard to dewater. Betz 1100 enhanced flocculation and improved the settling characteristics of the sludge better than did the other flocculants. The sludge contained approximately one-half of the ^{90}Sr and one-tenth of the ^{137}Cs.

The results of the scouting tests indicate that the traditional cold lime—soda-ash process is not adequate for this application. In addition to having the highest effluent hardness concentration, the process produces a large quantity of radioactive sludge since the calcium present in the lime also precipitates in the softening process. The remaining processes adequately softened the process wastewater; however, the caustic—soda-ash process minimized both the hardness and sodium in the effluent and produced an easily dewatered sludge.

Laboratory- and full-scale tests indicated that the iron did not improve softening under normal operating conditions when a sludge blanket was present, and it should be eliminated to improve the dewatering process. When the iron is removed from the scavenging precipitation process, the caustic process described above is obtained. However, contaminants, such as detergents, occasionally enter the process wastewater and inhibit the softening process. Tests indicated that under these conditions the presence of iron enhances the removal of calcium and magnesium. Therefore, the caustic—soda-ash and scavenging precipitation processes were used in ion exchange column tests. Since the presence of iron in the precipitation step will not affect the ion

Table 3. Results of Process Wastewater Softening Tests

	Concentration, mg/L			
	Lime—Soda-Ash Process	Caustic—Soda-Ash Process	Caustic Process	Scavenging-Precipitation Process
Chemical Requirements				
Na_2CO_3	190	95	0	0
$Ca(OH)_2$	125	0	0	0
NaOH	0	70	450	500
Alum	5	0	0	0
Iron	0	0	0	5
Betz 1100 Polymer	0.8	0.6	0.3	0.3
Effluent Characteristics				
Total Hardness[a]	60	49	14	10
Ca Hardness[a]	16	8	5	4
Na	83	80	260	290
Final pH	10.4	10.5	11.9	11.9
% Solids in Filter Cake	--	60[b]	50-60[b]	25[b]-50[c]

[a]Measured in mg/L as $CaCO_3$.

[b]No filter aid added to filter press feed.

[c]15 g/L diatomaceous earth filter aid added to filter press feed.

exchange performance, the caustic process was dropped from further consideration in small-scale ion exchange tests.

Ion Exchange Tests

The laboratory-scale ion exchange tests were performed in 1.3-cm (0.5-in.)-OD columns that contained 6.5 to 20 mL of material with 2.3 to 7.1 length-to-diameter ratios and 1- to 10-min residence times. Composite samples of the effluent were collected over 8- to 12-h periods and were analyzed for ^{90}Sr, ^{137}Cs, and gross beta activity. The zeolites and ion exchange resins that were tested in these small-scale tests are listed in Table 4.

Experimental breakthrough curves were obtained by plotting the mean throughput (measured in bed volumes) vs fractional breakthrough, based on the mean feed concentration for the run. These curves were linear when plotted on logarithmic-probability paper, indicating that the Thomas model is valid for these data. Therefore, volume-based distribution coefficients were approximated as the throughput in bed volumes at the 50% breakthrough point. Most experimental data included the 50% point; however, the experimental data for IRC-84 resin and CH zeolite materials using scavenging-precipitation softened water and CH using caustic—soda-ash water only went through 10 to 20% breakthrough. Their distribution coefficients were determined by extrapolation and are therefore more questionable than the remaining data. In column tests in which ^{137}Cs had not begun to break through at the time of shutdown, the K_d values are listed as greater than the total number of bed volumes which had passed through the column at that point.

The performance of the materials was initially evaluated by comparing these experimentally determined distribution coefficients (shown in Tables 5 and 6) for

Table 4. Sorption and Ion Exchange Materials Tested in the PWWS[a] Flowsheet

Material	Manufacturer	Costs ($/ft^3)	Description
Zeolon 400	Norton Chemicals	150	Clinoptilolite
Zeolon 500	Norton Chemicals	150	Natural chabazite-erionite mixture
Zeolon 700	Norton Chemicals	150	Ferriorite
Zeolon 900	Norton Chemicals	150	Synthetic mordenite
Linde 4A	Union Carbide	150	Inorganic zeolite
Linde A-51	Union Carbide	170	Inorganic zeolite
Ionsiv IE-95	Union Carbide	165	Synthetic chabazite
PDZ-140-D	Tenneco Speciality Minerals	15	Natural Na^+-rich clinoptilolite
PDZ-150-D	"	15	Natural K^+-rich clinoptilolite
PDZ-300	"	15	Natural chabazite
CH[b]	Chem Nuclear	5	Natural clinoptilolite
HCR-S	Dow Chemicals	67	Strong-acid resin
Amberlite IRC-84	Rohm & Haas	165	Weak-acid cation resin
Duolite CS-100	Diamond Shamrock	230	Weak-acid cation resin
Dowex TG-650-G12	Dow Chemicals	---	Strong-acid cation resin
Dowex XFS-43230	"	---	Radium-selective resin

[a]PWWS = Process Wastewater System.

[b]Sample obtained from Chem Nuclear Company.

Table 5. Experimentally Determined Cesium Distribution Coefficients[a]

Ion Exchange Material	Unsoftened Water[b,c]	Caustic–Soda-Ash Softened Water[b,d]	Scavenging-Precipitation Softened Water[b]
Zeolon 400	21,000	--	--
Zeolon 500	>15,000	--	--
Zeolon 700	>8,500	--	--
Zeolon 900	>14,000	--	--
Ionsiv IE-95	>15,000	>18,000	13,000[e]
PDZ-300	23,000	--	--
PDZ-140-D	>8,000	19,000	14,000[f]
CH	>2,000	15,000	10,500[f]
Linde 4A	--	8,400	3,400[e]
Linde A-51	3,600	--	--
TG-650-G12	1,000	--	--
XFS-43230	470	--	--
HCR-S	--	430	100[e]
IRC-84	--	400	160[f]
CS-100	--	500	3,600[e]

[a]Greater than indicates no breakthrough at the maximum throughput measured in bed volumes of water processed at shutdown.

[b]Calcium and total hardness data were obtained by wet chemistry analysis. Average values for the individual components do not sum to the total hardness due to inherent errors in the analytical techniques.

[c]Average cation concentrations of 50 mg/L Ca, 12 mg/L Mg, and 30 mg/L Na and total hardness of 150 mg/L as $CaCO_3$.

[d]Average cation concentrations of 5 mg/L Ca, 3 mg/L Mg, and 150 mg/L Na and total hardness of 27 mg/L as $CaCO_3$.

[e]Average cation concentrations of 5 mg/L Ca, 1 mg/L Mg, and 270 mg/L Na and total hardness of 18 mg/L as $CaCO_3$.

[f]Average cation concentrations of 1 mg/L Ca, 0.05 mg/L Mg, and 300 mg/L Na and total hardness of 4 mg/L as $CaCO_3$.

each water-softening process tested. The tests indicate that all of the zeolites are good sorbents for ^{137}Cs and that the chabazites (i.e., PDZ-300 and Ionsiv IE-95) had the highest capacity for ^{90}Sr removal from unsoftened process water (similar to that of the presently used HCR-S resin). In general, softening water prior to ion exchange decreases strontium and cesium sorption capacities of chabazites (due to the higher sodium concentration), while it increases the capacities of clinoptilolites (i.e., PDZ-140-D and CH), Linde A zeolites, and the resins for ^{90}Sr (which is still the limiting ion). Tables 5 and 6 show that there is a trade-off between increasing the sodium concentration and reducing the total hardness in the water. Most sorption materials have higher ^{90}Sr-loading capacities for caustic–soda-ash softened water (150 mg/L sodium and 27 mg/L total hardness) than for water treated by the scavenging precipitation process (300 mg/L sodium and 4 to 18 mg/L total hardness).

The exceptions to the above observation are the CH clinoptilolite and CS-100 and IRC-84 resins. Better performance would be expected for CS-100 with the high-pH scavenging-precipitation process, because the resin contains phenolic groups that ionize to remove cesium at pH 12. The increased strontium capacity for the CH zeolite with scavenging-precipitation softened water over caustic–soda-ash processed water does not agree with literature values or experimental data for similar materials. This value could have been incorrectly estimated because of

Table 6. Experimentally Determined Strontium Distribution Coefficients

Ion Exchange Material	Unsoftened Water[a,b]	Caustic—Soda-Ash Softened Water[b,c]	Scavenging-Precipitation Softened Water[b]
Zeolon 400	370	--	--
Zeolon 500	2,000	--	--
Zeolon 700	320	--	--
Zeolon 900	300	--	--
Ionsiv IE-95	2,200	3,800	1,000[d]
PDZ-300-17	3,100	--	--
PDZ-140-D	800	15,000	10,500[e]
CH	350	13,000[f]	36,000[e,f]
Linde 4A	--	13,000	12,000[d]
Linde A-51	800	--	--
TG-650-G12	1,500	--	--
XFS-43230	540	--	--
HCR-S	--	7,000	6,800[d]
IRC-84	--	15,000	31,000[e,f]
CS-100	--	500	3,800[d]

[a]Average cation concentrations of 50 mg/L Ca, 12 mg/L Mg, and 30 mg/L Na and total hardness of 150 mg/L as $CaCO_3$.

[b]Calcium and total hardness data were obtained by wet chemistry analysis. Average values for the individual components do not sum to the total hardness due to inherent errors in the analytical techniques.

[c]Average cation concentrations of 5 mg/L Ca, 3 mg/L Mg, and 150 mg/L Na and total hardness of 27 mg/L as $CaCO_3$.

[d]Average cation concentrations of 5 mg/L Ca, 1 mg/L Mg, and 270 mg/L Na and total hardness of 18 mg/L as $CaCO_3$.

[e]Average cation concentrations of 1 mg/L Ca, 0.05 mg/L Mg, and 300 mg/L Na and total hardness of 4 mg/L as $CaCO_3$.

[f]Distribution coefficients determined by extrapolation from the 10 to 20% breakthrough point.

inaccuracies associated with extrapolation of distribution coefficients from experimental data.

Amberlite IRC-84 has sorption capacities and kinetic properties that enable the resin to remove both monovalent and multivalent cations efficiently, but its affinity for divalent cations is much greater than its affinity for monovalent cations. Its sorption capacity is therefore more sensitive to the calcium ion concentration than to the sodium ion concentration, as indicated by the inverse relationship between the ^{90}Sr distribution coefficient and hardness. When the total hardness ranged between 18 to 27 mg/L, the column only loaded with divalent cations (indicated by negligible volume change during loading). When the hardness was lowered to 4 mg/L, the resin initially loaded with sodium ions, which resulted in degassing and swelling of the resin to approximately 210% of its original volume. In subsequent tests, this was avoided by conversion of the resin from its original hydrogen form to the sodium form prior to column loading. Degassing and swelling were not noticed in tests with any other sorption material.

Typical breakthrough curves for ^{90}Sr using unsoftened and softened feed are shown in Figures 1 and 2, respectively. All curves are typical of those obtained from small-scale column tests with low residence times. The data indicate that the model described

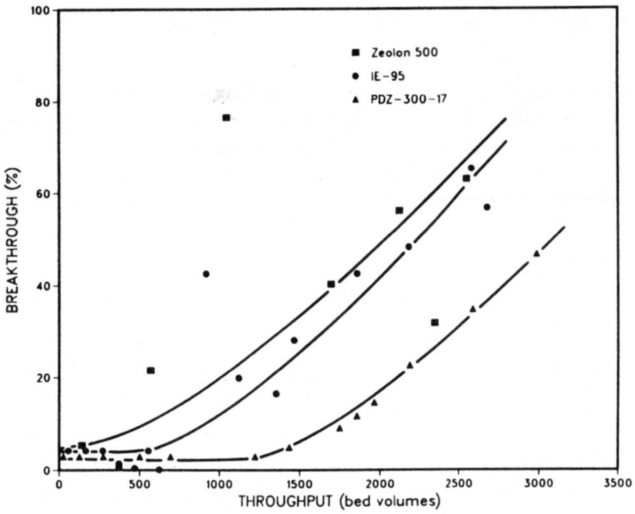

Figure 1. ^{90}Sr breakthrough curves for unsoftened process wastewater.

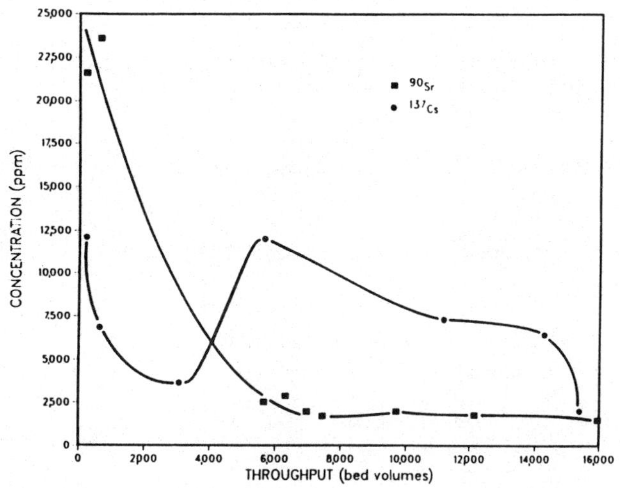

Figure 2. ^{90}Sr breakthrough curves for process wastewater softened by the scavenging precipitation process.

above should adequately predict resin performance. It is interesting to note that PDZ-300, the cheaper natural chabazite, had a higher ^{90}Sr capacity than did the synthetic material, IE-95. The Zeolon 500, an erionite-chabazite mixture, performed similarly to the synthetic material.

Variations of the ^{90}Sr and ^{137}Cs feed concentrations did not create operational problems as long as the feed concentration changed gradually during a run. Rapid changes in feed concentrations had detrimental effects on zeolites and strong-and weak-acid resins. Breakthrough curves such as those for IE-95 shown in Figure 3 were obtained when the composition of the feed stream exponentially decreased from 22,000 Bq/L ^{90}Sr and 12,000 Bq/L ^{137}Cs to 2000 and 1000 Bq/L, respectively, during the run. Premature breakthrough can be attributed in part to elution of the radionuclides from the zeolites by the feed solution when the concentration of the radionuclides in the feed fell below the equilibrium value created by the initially concentrated feed. It is presumed that the

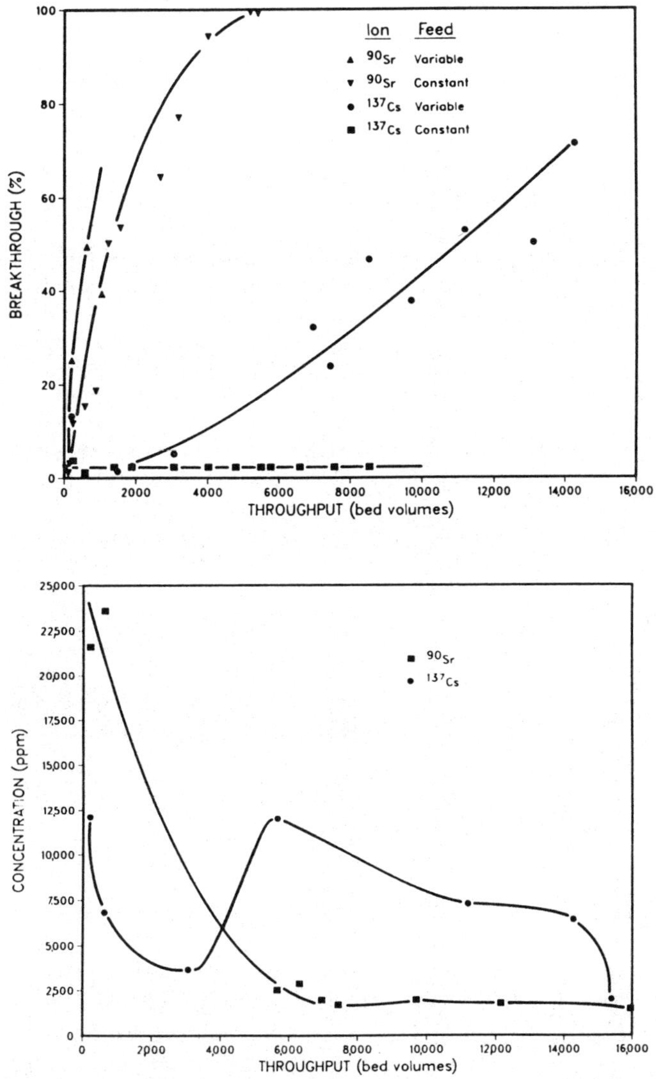

Figure 3. Effects of radionuclide feed concentration on Ionsiv IE-95 breakthrough curves. a) Breakthrough curves, b) Variable feed concentration; constant feed concentrations are 3500 Bq/L for ^{90}Sr and 440 Bq/L for ^{137}Cs.

concentration of other components in the feed also changed during this time frame and may have contributed to early breakthrough.

FLOWSHEET DEVELOPMENT

The alternative process flowsheets shown in Figures 4 to 7 were developed for fixed-bed ion exchange columns based on the results

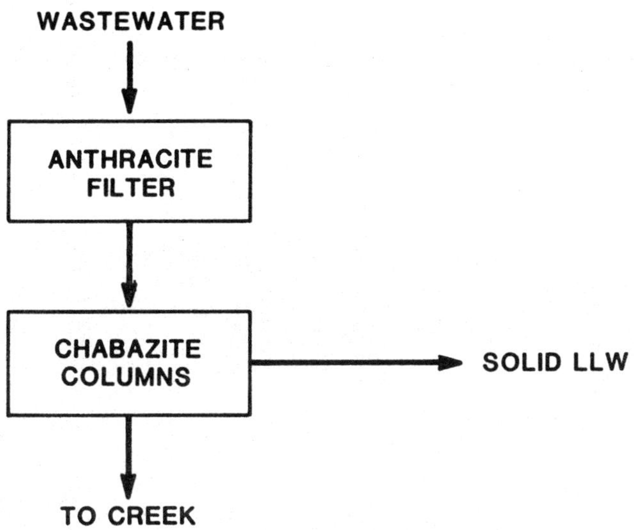

Figure 4. Proposed chabazite flowsheet for wastewater treatment.

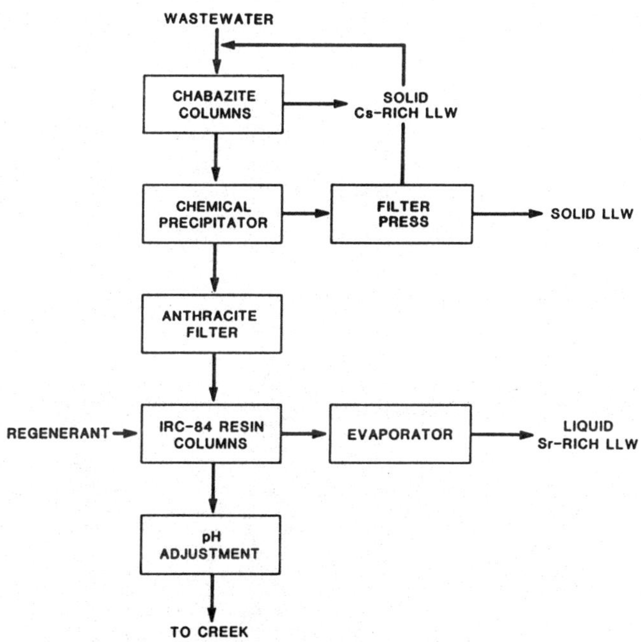

Figure 5. Proposed weak-acid resin flowsheet for wastewater treatment.

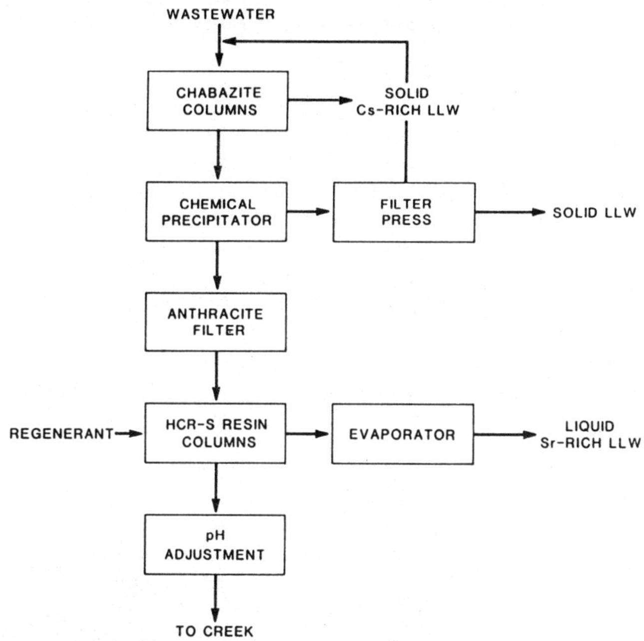

Figure 6. Proposed strong-acid resin flowsheet for wastewater treatment.

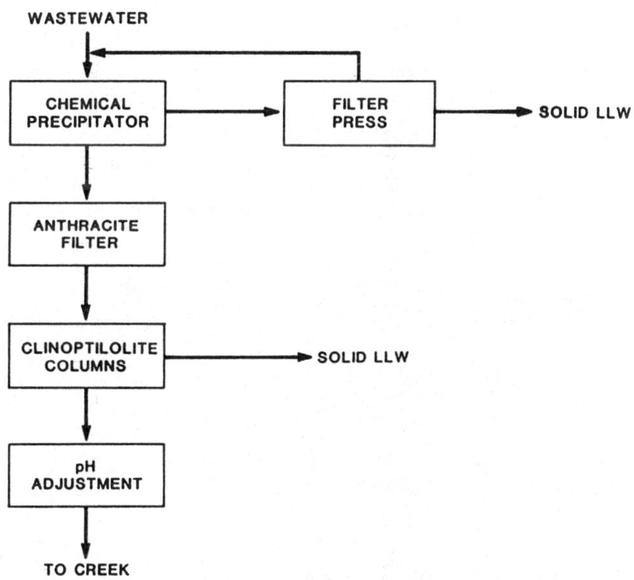

Figure 7. Proposed clinoptilolite flowsheet for wastewater treatment.

of the above scouting tests. The simplest proposed flowsheet, the chabazite flowsheet shown in Figure 4, uses the PDZ-300 chabazite to remove both cesium and strontium. Upon strontium breakthrough, the zeolites would be disposed of as solid low-level waste. If several zeolite columns were used in series, this process has the potential of being a simple, economical decontamination method that would produce only one type of solid waste.

The weak-acid resin flowsheet in Figure 5 utilizes the materials with the maximum sorption capacities for cesium and strontium, respectively (assuming the ^{90}Sr distribution coefficient for CH zeolite in Table 6 using scavenging-precipitation softened water is too high). It includes chabazite columns in series for cesium removal, followed by a caustic water softener for magnesium and calcium removal, and columns containing IRC-84 for strontium removal. Since laboratory tests have indicated that the use of IRC-84 may result in operational problems typically associated with weak-acid resins (i.e., swelling, degassing, and degrading), a similar flowsheet (the strong-acid flowsheet shown in Figure 6) was developed in which IRC-84 was replaced with HCR-S, a more forgiving strong-acid resin with a lower strontium-loading capacity which is more difficult to regenerate. The loaded zeolites would be disposed of as solid LLW. Two treatment options were considered for the loaded resins: regeneration and direct LLW disposal.

An alternative flowsheet (Figure 7) proposes that the PDZ-140 clinoptilolite be used to remove both cesium and strontium after calcium and magnesium have been removed from the feed by chemical precipitation. Although the cesium- and strontium-loading capacities on clinoptilolites are about one-half as large as those for the materials selected for use in the process illustrated in Figure 5, the flowsheet is significantly simplified and only one sorbent waste is generated. The process (as in that of Figure 4) also eliminates liquid LLW.

The above flowsheets were developed using experimental and full-scale data, when available, to predict the plant operating conditions at 9.45 L/s. An average feed composition of 4000 Bq/L ^{90}Sr and 400 Bq/L ^{137}Cs was used. Maximum effluent concentrations were limited to 10 Bq/L for ^{90}Sr and 40 Bq/L for ^{137}Cs. These values were based on the respective 10 CFR 20 limits of 11.1 and 740 Bq/L (Office of the Federal Register, 1981) for each individual ion. The limit for mixtures states that the summation of the fractional releases should not exceed 1. The flowsheets were developed for a train of four 3.7-m^3 columns in series for the nonregenerable sorbents, while two 1.34-m^3 columns run in parallel were assumed for resins which would be regenerated (as presently exist in the Process Wastewater System). The mean residence time through each column would be 6.5 min for 3.7-m^3 columns and 4.7 min for the 1.34-m^3 columns.

The experimental breakthrough data were fitted by means of the model described above, assuming R = 1. The parameters N and T were determined for each set of experimental data by iteration using Equations (5) and (6) (Tan, 1984). Most of the curves are approximately linear when plotted on probability paper, and the models fit the data well. A linear relationship between column length and residence time was used to predict the performance of regenerable columns at full-scale conditions. Additional calculations were needed to determine the replacement rate of nonregenerable columns.

In order to maintain the desired effluent concentrations in each series of nonregenerable columns used in the flowsheets, the operating method would need to include removing the first column in the series at some point, moving the remaining columns one position forward (countercurrent to the water flow), and placing a fresh column in the last position in the series. Repetitive cycles of this type of operation were modeled using a numerical analysis solution of the Thomas equation developed by R. M. Wallace (Collins et al., 1986).

Calculations of this type were made for each set of nonregenerable columns in Figures 4 to 7. The replacement frequency of the first column in the series was set at the value required to maintain the maximum effluent concentration (at the end of a cycle) equal to the limiting ^{90}Sr concentration. The resulting frequencies of column regeneration or replacement are shown in Table 7, along with the total amount of waste that would be generated under these conditions. The generation rates of liquid LLW were based on actual plant experience, but the sludge generation was based on both laboratory- and full-scale data. The data for the two previous treatment processes used in the Process Wastewater System are included in Table 7 for comparison.

The results indicate that the chabazite, weak-acid resin, and clinoptilolite flowsheets are significant improvements over the previous processes, while the strong-acid resin flowsheet produces approximately the same amount of waste as the SPIX process. Slightly less total waste would be produced by not regenerating the IRC-84 resin in the

Table 7. Waste Generation By Proposed Flowsheets

Flowsheet	Flowsheet Description			Regeneration or Replacement Rate, bv[a]		Waste Generation Rate, L/L Water Processed				
	Presoftening Sorbent	Clarifier Type	Postsoftening Sorbent	Presoftening Sorbent	Postsoftening Sorbent	Presoftening Sorbent	Sludge	Postsoftening Sorbent	Liquid Low-Level Waste	Total Wastes
Chabazite	PDZ-300	None		4850	0	2.08E-04	0.00E+00	0.00E+00	0.00E+00	2.08E-04
Weak-Acid Resin[b]	PDZ-300	Caustic[c]	IRC-84	23800	12500	4.22E-05	1.71E-04	4.74E-06	4.17E-05	2.60E-04
Weak-Acid Resin[d]	PDZ-300	Caustic	IRC-84	23800	30600	4.22E-05	1.71E-04	3.32E-05	0.00E+00	2.46E-04
Strong-Acid Resin[b]	PDZ-300	Scav. Prec.[e]	HCR-S	23800	3260	4.22E-05	2.85E-04	9.01E-06	7.02E-05	4.06E-04
Clinoptilolite		Caus. Soda[f]	PDZ-140	0	13750	0.00E+00	1.71E-04	7.31E-05	0.00E+00	2.44E-04
SPIX[g]		Scav. Prec.	CS-100	0	2000	0.00E+00	2.85E-04	3.23E-05	5.69E-05	3.74E-04
FIX[h]		Scav. Prec.	HCR-S	0	400	0.00E+00	0.00E+00	9.01E-05	6.17E-04	7.07E-04

[a]bv = bed volumes.

[b]Resin regenerated.

[c]Caustic = scav. prec. without iron.

[d]Resin disposed of as solic waste.

[e]Scav. prec. = scavenging precipitation softening process

[f]Caus. soda = caustic-soda ash process.

[g]SPIX = scavenging-precipitation ion exchange.

[h]FIX = filtration-ion exchange.

weak-acid resin flowsheet. The chabazite flowsheet generated the least amount of waste, but the wastes generated by the chabazite, weak-acid resin, and clinoptilolite flowsheets are so similar that pilot-scale testing and economic analyses will be required to determine which process should ultimately be used.

Note that the flowsheet that generates the lowest amount of total waste utilizes a chabazite zeolite which has a relatively low distribution coefficient for ^{90}Sr. The sludge generated in the other flowsheets is 1 to 2 orders of magnitude greater than the amount of waste sorption material and offsets gains made by increased resin efficiency. These data also indicate that errors due to estimation of resin performance should have little bearing on the overall outcome of flowsheet analyses.

LARGE-SCALE TESTING

The strong-acid flowsheet shown in Figure 6 has been implemented at full-scale at ORNL using existing equipment from the previous SPIX (Figure 7) process. The scavenging-precipitation softening process is being used because the precipitator design requires a fluffy floc for proper operation. Sludge is dewatered to 50% solids using a plate and frame filter press with a diatomaceous earth filter aid and is stored for permanent disposal. The resin columns are run until calcium begins to break through the column, which occurs prior to the ^{90}Sr breakthrough. The columns are loaded with ~4.4E-10 g ^{90}Sr per g resin at that point. This process has reduced the total waste generation rate from 184 m^3/year of liquid LLW to 122 m^3/year of LLW, of which only 23 m^3/year is liquid waste.

The chabazite flowsheet shown in Figure 4 is also being tested at pilot- and full-scale operations (Collins et al., 1987). Skid-mounted equipment is being leased from the Chem Nuclear Company to test this flowsheet and to develop techniques for operating a series of sluicible columns. Two columns, each containing 3.7 m^3 of Ionsiv IE-95 zeolite, were used in series to treat 6700 m^3 (1810 bv) of process wastewater until ^{90}Sr effluent concentrations reached 10 Bq/L, yielding an instantaneous decontamination factor of 400. No ^{137}Cs breakthrough was detected at this point. The columns were loaded with 4.1E-10 g ^{90}Sr and 3.7E-10 g ^{137}Cs per g of resin at shutdown. These data confirmed the efficiency for ^{90}Sr removal (distribution coefficient of 1700) that had been obtained in the small-column tests.

Since the zeolite was not loaded to capacity at shutdown, a more efficient flowsheet for Figure 4, consisting of a series of smaller columns, is being tested. The 10% plant-scale equipment contained the inexpensive natural chabazite, PDZ-300. The performance with this sorbent (^{90}Sr distribution coefficient of 6500) is much better than that obtained with IE-95 and is better than the predicted values based on laboratory-scale data. This improved performance may be due to slower diffusion through the natural material and indicates that equilibrium may not have been reached in the small-scale column tests using PDZ-300.

Tests are planned to evaluate the other promising sorbents and to develop treatment and packaging processes of the spent sorbents for permanent disposal. The zeolites are a nonhazardous aluminosilicate clay that can be treated with heat to reduce the volume by a factor of ~2 and the leachability of the ^{90}Sr and ^{137}Cs (Kultgren et al., 1981). The sludge can then easily be solidified in concrete (Vanura and Konecny, 1984) for disposal.

SUMMARY

Improved treatment processes are being developed at Oak Ridge National Laboratory to treat slightly contaminated process wastewater. A number of zeolites have been found to be excellent sorption materials (with distribution coefficients up to 23,000) for ^{137}Cs from unsoftened process water. Chabazites have the best sorption capacity for ^{90}Sr under the same conditions, but the kinetics are relatively slow. Treatment with chabazites alone would require the use of a number of columns operated in series. Several materials prove to be good sorbents for ^{90}Sr when the feed has been softened to remove calcium and magnesium ions. For example, clinoptilolites and weak-acid resins have loading capacities of 10,000 to 25,000 bed volumes.

Improvements made at Oak Ridge National Laboratory, based on these data, have already reduced the LLW generated by treatment of process wastewater to 66% of the original volume. Proposed processing methods could eliminate all liquid LLW and reduce the solid waste by an additional one-third.

ACKNOWLEDGMENT

The assistance and support of the personnel in the Waste Management Operations Group of the Operations Division at Oak Ridge National Laboratory are greatly appreciated. Research was performed at Oak Ridge National Laboratory, which is operated by Martin Marietta Energy Systems, Inc., under contract DE-AC05-84OR21400 with the U.S. Department of Energy.

NOTATION

C – concentration of solute in effluent, g-mol/cm^3.
C_o – concentration of solute in feed, g-mol/cm^3.
f – flow rate through column, cm^3/s.
K_a – mass transfer coefficient, 1/s.
K_d – distribution coefficient, dimensionless.
K_d' – distribution coefficient, cm^3/g.
N – number of mass transfer units, dimensionless.
q – concentration of solute in solid, g-mol/g.
q^* – concentration of solute in solid at equilibrium, g-mol/g.
q_v – concentration of solute per unit volume of sorbent bed, g-mol/cm^3.
R – separation factor, dimensionless.
T – throughput parameter, dimensionless.
V – effluent volume, cm^3.
v – sorbent bed volume, cm^3.
X – concentration of solute in fluid phase, dimensionless.
Y – concentration of solute in solid phase, dimensionless.

Greek Letters
ρ_b – bulk density of sorbent, g/cm^3.

LITERATURE CITED

Browder, F. N., "Radioactive Waste Management Oak Ridge National Laboratory," Oak Ridge National Laboratory Report No. 2601, Oak Ridge, TN (1959).

Chilton, J. M., and L. C. Lasher, "Decontamination of Low-Level Liquid Waste at Oak Ridge National Laboratory using a Scavenging-Precipitation, Ion Exchange Process," in Waste Isolation in the US and Elsewhere, Technical Programs and Public Communications. Vol. 2 Low-Level Waste, University of Tucson, Tucson, AZ (1982).

Collins, E. D., J. M. Begovich, C. H. Brown, D. O. Campbell, L. C. Lasher, M. I. Morris, S. M. Robinson, and C. B. Scott, "An Improved Ion Exchange Method for Treatment of Slightly Contaminated Waste Waters," presented at the American Nuclear Society International Meeting on Low-, Intermediate-, and High-Level Waste Management and Decontamination and Decommissioning, Niagara Falls, NY (1986). (Proceedings to be published in 1987).

Collins, E. D., D. O. Campbell, L. J. King, J. B. Knauer, and R. M. Wallace, "Development of the Flowsheet Used for Decontaminating High-Activity-Level Water," in The Three Mile Island Accident, Diagnosis and Prognosis, ACS Symposium Series, 293, American Chemical Society, Washington, DC, 212 (1986).

Foust, S. D., and O. M. Aly, Chemistry of Water Treatment, Butterworth & Co., Ltd., London (1983).

Hamer, P., J. Jackson, and E. F. Thurston, Industrial Water Treatment Practice, Butterworth & Co., Ltd., London (1961).

Helfferich, F. G., Angew. Chem. Int. Ed. Engl., 1(8), 440 (1962).

Hiester, N. K., and T. Vermeulen, Chem. Eng. Prog., 48, 505 (1952).

Hiester, N. K., T. Vermeulen, and G. Klein, in Chemical Engineers' Handbook, 5th Ed., Perry, J. H., et al. (Eds.), McGraw-Hill, New York (1973).

Kultgren, A., et al., "The PILO Process: Zeolite and Titanates in the Treatment of Spent Ion Exchange Resins," presented at the Symposium on Management of Radioactive Wastes from Nuclear Power Plants, IAEA SR-56/17, Karlsruhe, FRG (1981).

Mercer, B. M., "Ion Exchange Equilibria of Trace Cesium-137 and Strontium-85 in Multicomponent Systems," Battelle-Northwest Lab. BNWL-SA-1173, Conf-670401-9, Richland, WA (1967).

Office of Federal Register, National Archives and Records Service, General Services Administration, Code of Federal Regulations, Part 10, U.S. Government Printing Office, Washington, DC (1981).

Tan, H. K. S., Chem. Engr., 91(26), 57 (1984).

Vanura, P., and Konecny, C., Nukleon, 4, 9 (1984).

DESIGN AND TESTING OF NATURAL/BLENDED ZEOLITE ION EXCHANGE COLUMNS AT WEST VALLEY

P. Burn, D.K. Ploetz and A.K. Saha ■ West Valley Nuclear Services, Co., Inc., P.O. Box 191, West Valley, NY 14171

D.C. Grant and M.C. Skriba ■ Westinghouse Research and Development Center, 1310 Beulah Road, Pittsburgh, PA 15235

A zeolite ion exchange system has been designed and fabricated for decontaminating various radioactive liquid wastes being generated during the decommissioning activities at the former nuclear fuel reprocessing plant in West Valley, New York. Variables affecting the column design and performance were investigated in the laboratory. Results of testing showed that the removal of the radioactive contaminants is strongly dependent upon waste composition and residence time. The use of a zeolite bed containing both a natural zeolite (phillipsite) and a processed zeolite (Linde A-51) were found to successfully handle the waste streams.

West Valley Nuclear Services Company, a wholly owned subsidiary of Westinghouse Electric Corporation, is conducting various nuclear waste processing and decommissioning operations under a contract for the U.S. Department of Energy. Included in these operations is the development of an integrated liquid waste processing system. This system involves the application of ion exchange technology for removing radioisotopes (i.e., cesium and strontium) from the various waste streams being generated. After extensive laboratory testing, phillipsite, a naturally occurring orthorhombic zeolite, has been selected as the reference ion exchange media. For some specific applications, Linde A-51, a type "A" strontium selective zeolite, will be blended with phillipsite. This paper discusses the design of natural/blended zeolite ion exchange columns with respect to the variables affecting column performance. Cold testing of the columns are currently in progress. Extensive testing has been performed in a 1/12 scale model to determine various flow characteristics.

VARIABLES AFFECTING COLUMN DESIGN

The column length, diameter, flow distributors, and bottom discharge connections are some of the important variables for the zeolite column design. The total exchange capacity available and the pressure drop across the bed are both directly proportional to the zeolite bed length. The larger the column diameter the greater the liquid hold-up in the pores of the bed, and thus the higher the residence time per length of bed. Increasing the ratio of the length to diameter (i.e., aspect ratio) can improve the utilization of the bed if the contaminant loading is favorable (i.e., sharp breakthrough curve), but has little effect if the loading is unfavorable. Due to the relatively high capacity of the zeolite bed and thus minimal contaminant breakthrough, the laboratory testing conducted thus far has given no indication of whether or not a high aspect ratio is desired. A relatively small aspect ratio (i.e., 1.25) has been chosen in the full-scale design to provide a high residence time and capacity for a given length of bed and to minimize the total pressure drop.

The design of the system must also address the fact that zeolites are a more friable material than organic resins, and as such, produce more fines during the course of operation. Zeolite ion exchange columns thus require more frequent backwashing and bed preparation. Provisions must be made in the design to allow removal of the fines from the bed by backwashing. In addition it is advisable to keep the zeolite bed flooded to help minimize physical deterioration of the zeolite.

Figure 1 shows a schematic of the zeolite column to be used at West Valley. Two identical columns have been designed and fabricated. The columns are fitted with "Johnson" screen laterals for feeding and withdrawing effluents at the top and at the bottom. A "hanging basket" type Johnson screen assembly is provided at the top to remove fines during bed preparation.

The design of the flow distributor, shown in Figure 1, is important for achieving uniform liquid distribution over the bed surface and thus to minimize channeling within the relatively short zeolite bed (~ 1.5 m). The "Johnson" screen laterals are inserted within tubes having equidistance perforated holes facing downwards for the upper distributor and upwards for the lower distributor. Liquid trickles across the length of the bed and is withdrawn by the "Johnson screen" laterals at the bottom.

Zeolites do have a higher tendency to pack when used for an extended period of time under enhanced fluid pressure. Due to this packing tendency, the design of the column bottom discharge connections/valving is also very important. The discharge nozzle has been placed directly at the dead center of the bottom dish and is fitted with a specially designed flush bottom tank valve to facilitate complete straight thru removal of spent zeolite slurries.

PROCESS VARIABLES AFFECTING COLUMN DESIGN AND PERFORMANCE

The results of small-scale column tests being conducted are summarized in Figures 2 through 4. The testing (1) has shown that isotope (Cs-137 and Sr-90) removal from the waste streams is strongly dependent upon waste composition and residence time. Even seemingly slight variations in waste compositions resulted in significantly different isotope loading/separation characteristics (1). The solution pH also seemed to play a role, particularly for the decontamination of Sr-90 isotopes. The relative slow loading kinetics of zeolites result in the need for relatively long residence times.

Two simulated waste streams have been tested. One waste stream, referred to as "supernatant", consists of a mixture of various sodium salts, (e.g., $NaNO_3$, $NaNO_2$, Na_2SO_4, Na_2CO_3, NaCl, etc.) to produce a final concentration of about 200 ppm TDS (total dissolved solids). The other waste stream is a neutralized nitric acid waste which has a final concentration of about 5000 ppm TDS. Cesium and strontium contaminants are added to make the wastes radioactive. The waste streams have been tested on two separate beds of the phillipsite zeolite. As shown in Figure 2, after thousands of bed volumes, the column treating the supernatant still produced the desired decontamination factors, (DF) (i.e., 500-1000) for both the cesium and strontium isotopes. The contaminants have not shown any sign of significant breakthrough.

From the seemingly analogous chemical and isotopic composition of the two waste streams, it was expected that the second stream consisting of only the neutralized nitric acid (i.e., sodium nitrate) would also be successfully processed with the phillipsite bed for thousands of bed volumes and produce required DF's. However, as shown in Figure 2, it was found that while the cesium DF was comparable to the first waste stream, the strontium broke through very rapidly. In fact, the strontium broke through completely (i.e., DF=1) in less than 100 bed volumes. The strontium DF could not be improved by varying the pH (7 to 9) (see Figure 3) or by increasing the residence time from 20 to 40 minutes.

In order to improve the strontium DF, a strontium specific zeolite (Linde A-51) was blended with phillipsite in a volume ratio of phillipsite to A-51 of 2. Both waste streams were tested with this mixture in a manner identical to that used with phillipsite alone. For the supernatant waste stream, it was found that there was no significant improvement in terms of Cs/Sr DF's, even after increasing the residence time from 20 to 40 minutes. Thus, for this waste stream, the mixed bed zeolite did not offer any significant advantage over the phillipsite alone.

As shown in Figure 4, for the neutralized nitric acid stream, the mixed bed zeolite produced a dramatic increase in the strontium DF as well as an improvement, although smaller, in the cesium DF. In addition, the effect of residence time was profound using this mixed bed on this waste stream. A residence time of 20 minutes failed to achieve proper DF's; whereas, a residence time of 40 minutes resulted in the desired DF's being exceeded.

ZEOLITE BED FLOW CHARACTERISTICS

Tests have been performed in a 1/12 scale model to determine various flow characteristics across the zeolite bed. Bed pressure drop (ΔP), flow flux for bed fluidization, optimum velocity for backwashing and fines removal are the parameters measured. From the test data it appears that insignificant ΔPs can be expected across the full scale zeolite bed. Various flux (gpm/ft^2) rates have been tested in the scale model to determine what rate is required for desired fluidization. A flux in the range of 4-6 gpm/ft^2 failed to lift the bed properly to cause desired fluidization (e.g., air and fines removal). An increase in flux to the range of between 7.5 and 8.5 gpm/ft^2 showed improvement and caused moderate fluidization. A further increase in flux to the range of 10.5 to 11.5 gpm/ft^2 resulted in the bed being perfectly fluidized.

DISCUSSION

Extensive testing in the laboratory scale and 1/12 scale model indicates that the zeolite ion exchange columns should be able to meet the process requirements. Phillipsite alone, or blended with A-51, possesses very high capacity for both cesium and strontium isotopes. Care has been taken in the column mechanical design to assure uniform flow distribution and sufficient residence time to overcome slower diffusional kinetics common to zeolitic ion exchange. A "goose-neck" piping is provided at the column exit to keep it flooded.

The seemingly anomalous behavior of the strontium ion exchange in the neutralized nitric acid waste stream is not fully understood. One possible explanation could be that in the supernatant stream there are ions which may interact with the strontium and make it more amenable to exchange onto the zeolite. The neutralized nitric acid waste, which is basically a very dilute solution of sodium nitrate, may be forming a complex ion of a particular radius which prevents its exchange, and thus simply passes through the phillipsite bed without any active exchange.

ACKNOWLEDGEMENT

This work was performed under U.S. Department of Energy Contract No. DE-AC07-81-NE44139.

LITERATURE CITED

1. Grant, D. C., M. C. Skriba, D. K. Ploetz and A. K. Saha, "Effect of Process Variables on the Removal of Contaminants from Radioactive Waste Streams Using Zeolites," presented at "Waste Management '87," University of Arizona/ASME/ANS, Tucson, AR, March 1987.

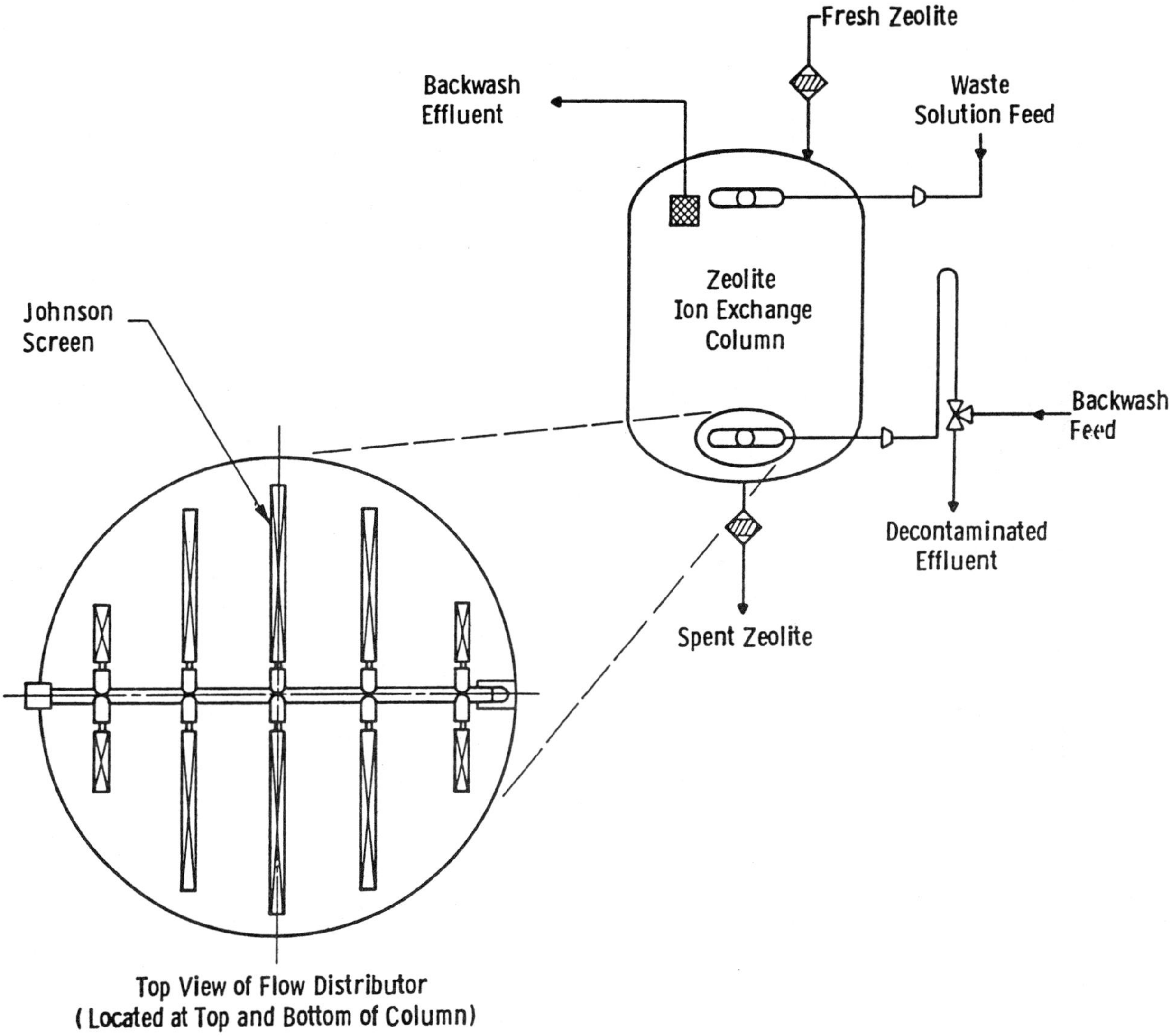

Figure 1. Schematic of zeolite ion exchange column.

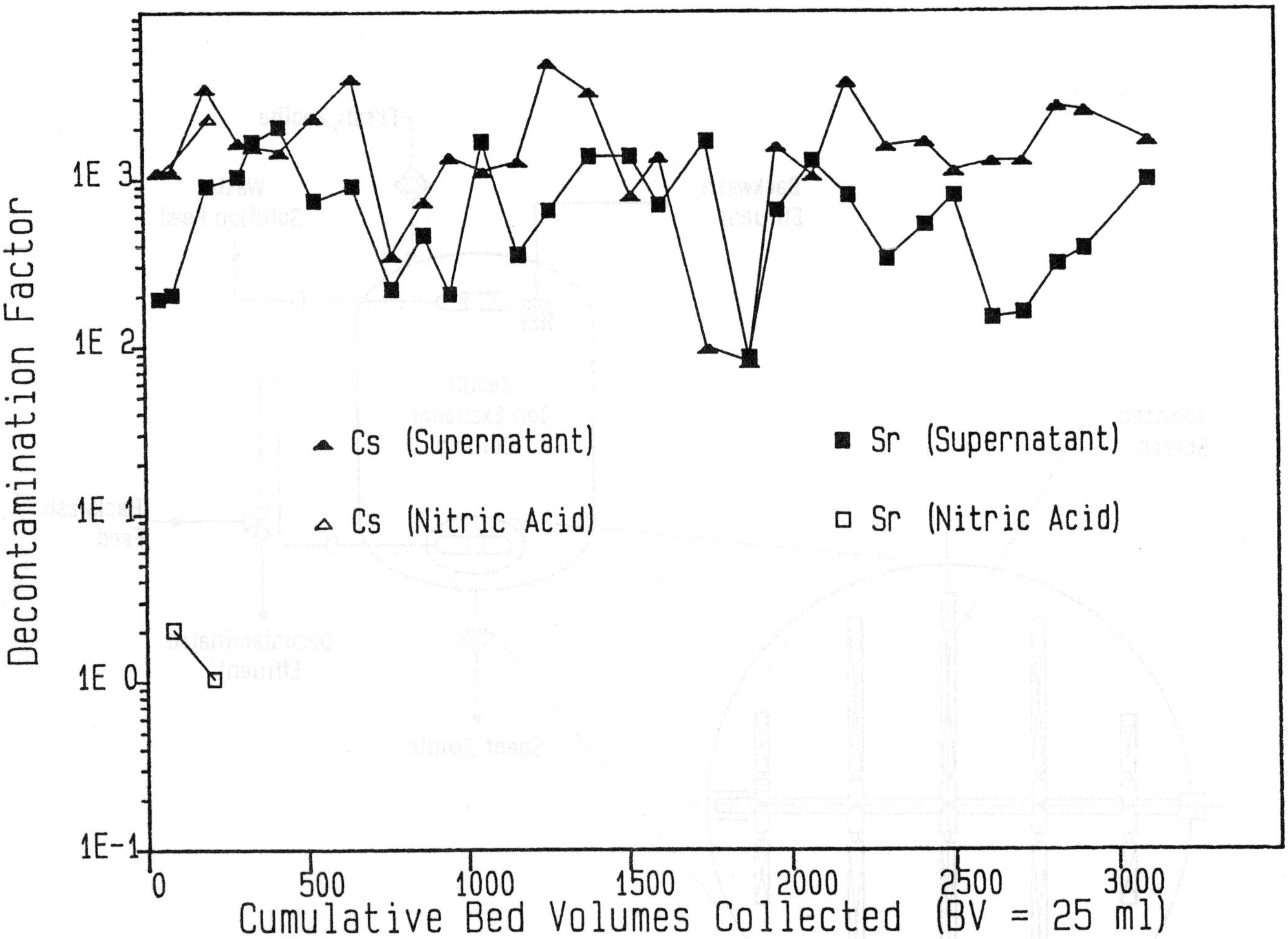

Figure 2. Supernatant and nitric acid waste treatment using phillipsite.

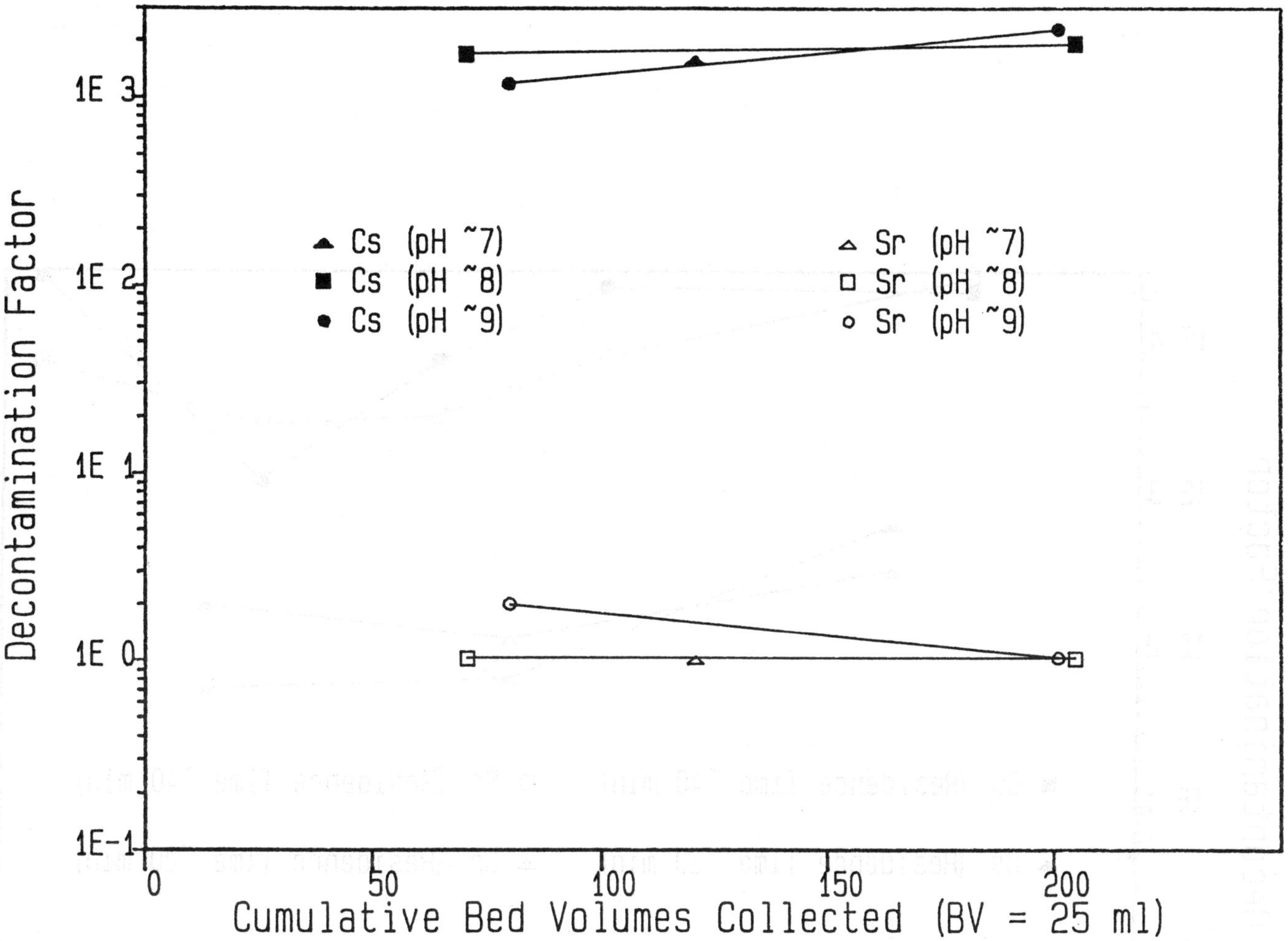

Figure 3. Effect of feed pH on nitric acid waste treatment using phillipsite.

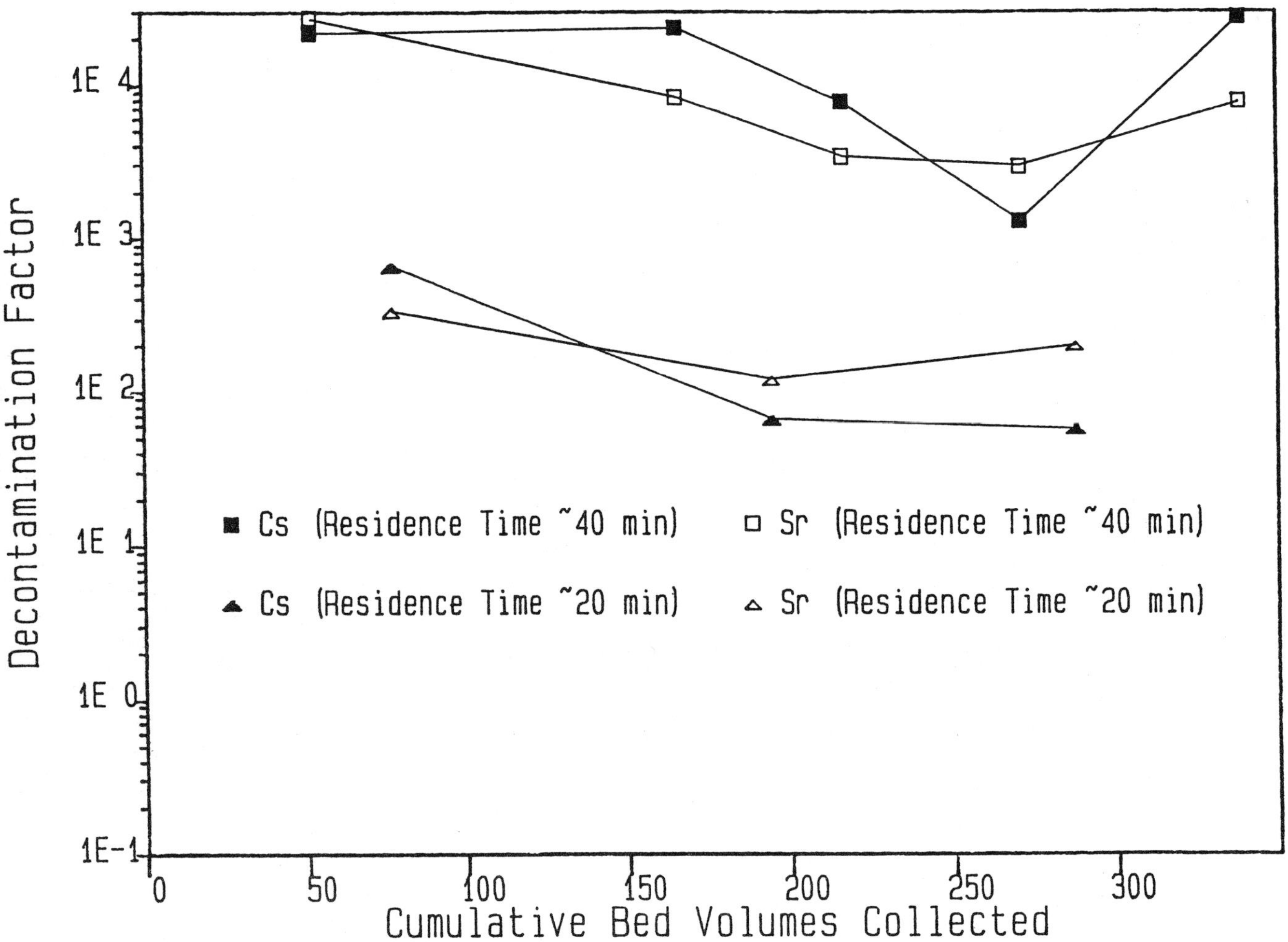

Figure 4. Effect of residence time on nitric acid waste treatment using phillipsite/A-51 mixture.

TREATMENT OF VARIOUS NUCLEAR GENERATING STATION WASTE WATERS WITH DURASIL™ MEDIA

Aaron Barkatt, Karen A. Michael and Hamid Hojaji ■ Vitreous State Laboratory, The Catholic University of America, Washington, DC 20064
Herbert G. Sutter ■ Duratek Corporation, 6411 Ivy Lane, Greenbelt, MD 20070

The development of selective sorption and ion exchange media based on porous silicate glass and charcoal for nuclear plant waste-water decontamination is described. This development involved laboratory studies with non-radioactive and radioactive dopants, laboratory tests on actual plant water samples, in-plant tests and full-scale operations. Development and testing are detailed in the cases of two PWR plants. The introduction of selective media yields significant improvements in throughput in both cases.

In order to meet the needs of the nuclear power industry for more effective volume reduction in wastewater treatment, the Vitreous State Laboratory at The Catholic University of America, with the support of Duratek Corporation, has undertaken a project of developing novel, highly selective ion exchange media to remove solute species which contribute significantly to the radioactivity of such wastewater. The high selectivity of these media in removing species such as cesium, iodine and cobalt has made it possible to obtain much higher capacities and bed lifetimes than are achieved by ion-exchange resin technology. The active sites on these ion exchange sorption media either do not sorb chemically similar solutes present at much higher concentrations in the water, e.g. sodium, chloride and calcium, or sorb such species in a form which permits a subsequent rapid and highly favored displacement by cesium iodide and cobalt, respectively. Other solutes which contribute to a stream radioactivity, such as manganese, chromium, silver and strontium can also be preferentially removed using such novel selective media. Several selective media of this type, marketed by Duratek Corporation under the trademark Durasil™, have been introduced on a large scale in the decontamination of aqueous waste streams in light water reactor (LWR) nuclear power plants.(1)

Both the major element chemistry and the identity of the major radioactive contaminants in aqueous LWR streams exhibit considerable differences among various plants as well as among various streams in a single plant. For instance, in cases where leakage from fuel elements is minimized the main contaminants are formed in the activation of corrosion products, and they include radio-isotopes of Fe, Cr, Mn and, in particular, Co. In cases where a substantial amount of leakage from the fuel elements takes place the presence of Ca and I becomes a major problem. However, if the water is stored in hold-up tanks for several weeks before treatment and disposal, iodine activity, which is largely due to I-131 (8-day half-life) decays away and Cs radioisotopes (Cs-137, 134) become the predominant sources of activity. The major element chemistry also exhibits wide variations, e.g. between low-conductivity streams such as condensates and primary coolant and high-conductivity streams such as service water in plants using saline service water, water used in laundry, clean-up operation, etc., and solutions resulting from resin regeneration. In almost every case the use of highly selective ion exchange and sorption media can result in considerable improvement in capacity and, consequently, in volume reduction factors, compared with the use of conventional broad-spectrum ion exchange resins. However, because of the large differences among various streams, different selective ion exchange or sorption media or combinations of such media have to be selected and optimized for each type of stream encountered in plant operation.

This process includes extensive laboratory testing and small-scale on-site testing prior to full-scale introduction of the optimized water purification system in plant operations. In general this process consists of five stages:

Stage (i)

Laboratory scale non-radioactive tests on simulated streams: These tests are used for preliminary screening of available ion exchange and sorption media to identify those which give the best performance in terms of capacities and decontamination factors in an environment reflecting both the major element and trace element chemistry of a particular stream in the nuclear industry. The simulated streams are formulated on the basis of a comprehensive analysis of samples of actual plant water using dc plasma spectroscopy and atomic absorption to determine anions, gamma spectroscopy, pH and conductivity measurements, and, whenever required, additional measurements, e.g. of organic solutes and particulate matter. The column size used in these laboratory tests is 0.3 to 15 mL, typically around 1 mL. In cases where available bed materials do not give satisfactory performance, attempts are made to modify existing media or develop new media and combinations of media to obtain improved capacities and decontamination factors. At this stage, tests are carried out with non-radioactive simulated solutions doped with parts-per-million levels of cesium, iodide, cobalt, etc., which can be analyzed using conventional spectroscopic methods.

Stage (ii)

Laboratory scale radioactive tests on simulated streams: At this stage, the simulated solutions are doped with parts-per-million levels of cesium, iodide, cobalt, etc. which can be analyzed using realistic levels (usually around 10^{-3} μCi/L) of Cs-137, I-131, Co-60, etc., and column performance is monitored by means of gamma spectroscopy. These tests permit much more quantitative evaluation of column capacities, since the high dopant concentrations necessary for chemical analysis in stage (i) usually result in much smaller column capacities than those obtained at the trace levels at which the corresponding elements are present in reactor streams. Another objective of the radioactive tracer tests is realistic evaluation of decontamination factors, since the relatively poor sensitivity of conventional chemical methods used in stage (i) prevents quantitative determination of high decontamination factors.

Stage (iii)

Laboratory scale tests on actual plant water: Tests are carried out on actual samples of plant water using media found to be promising in stages (i) and (ii) compared against media in current plant use for a particular stream. The column sizes are similar to those used in stages (i) and (ii).

Stage (iv)

Intermediate scale on-site tests: These tests are conducted with typical bed volumes of 0.1 to 5 liters, comparing the performance of optimized media selected in stages (i), (ii) and (iii) with the performance of media in current use. These tests are also used to identify optimum operating conditions, in particular the residence time.

Stage (v)

Full-scale plant operations based on the optimization of column media and operating conditions in the course of the previous stages.

Several case histories illustrating this approach are described below.

SALEM

The Salem Nuclear Generating Station in Salem, New Jersey, is a pressurized water reactor plant owned by Public Service Electric and Gas Company. The service water used in this plant is obtained from Delaware Bay and is highly saline. The levels of activated corrosion products, especially cobalt radioisotoptes, which eventually reach the aqueous waste streams, are considerable. On the other hand, fuel integrity is high, resulting in low concentrations of cesium and iodine. Due to the high salinity and high conductivity of the incoming service water, the conductivity of the waste stream is high, amounting on the average approximately 1500 μho·cm^{-1}, even though some of the water used in the plant undergoes de-ionization prior to use. At times, conductivities as high as 7000 μmho·cm^{-1}, are observed. The high salinities complicate the removal of cobalt by ion exchange, particularly since significant amounts of divalent cations (Mg^{2+}, Ca^{2+}) are present in the diluted sea-water streams, as well as high levels of complexing anions such as chloride and, in particular, sulfate.[2]

Stage (i)

A total of 183 column runs on simulated

solutions doped with 1 to 5 mg/L of non-radioactive cobalt were conducted in order to identify and optimize ion exchange or sorption media suitable for the removal of cobalt from highly saline aqueous streams. At the beginning of these strdies it was found that, as expected, the capacity of Nuclear Grade cation exchangers dropped sharply from 45 g/L Co in de-ionized water to 0.4 g/L in 20% sea-water and 0.1 g/L in full-strength sea-water.[1] The stage (i) experiments resulted in the development and optimization of a selective medium for cobalt removal, designated Durasil™ 70.[1,2] This is a cation exchanger based on charcoal with a particle size of -50+100 or -50+200 mesh (slightly smaller than the -20+50 mesh size of a typical organic resin which was used in the comparative evaluation of Durasil™ 70. The density of wet Durasil™ is approximately 1.1, similar to that of organic resins. Durasil™ 70 can be used at temperatures up to 60°C and over the pH range between 2 and 12 even in highly saline aqueous solutions. Its capacity for cobalt sorption drops slightly from 3.0 g/L Co in de-ionized water to 2.1 g/L in 8% sea-water but then it begins to rise again, reaching 4.4 g/L in full-strength sea-water. In addition to cobalt, it can also effectively remove from aqueous solutions other metals which contribute in certain cases to radioactive contamination, such as Mn, Cr, Ag and Cu. The sorption process is thought to be the result of complex formation between cobalt and organic species on the surface of the medium. The complicated dependence of cobalt sorption capacity on sea-water content is apparently related to changes in cobalt speciation. At higher sea-water concentrations the predominant forms of cobalt are sulfate and chloride complexes rather than the aquo form of the cobalt cation.[2] The observation that capacity in the case of Durasil™ is not inversely related to ionic strength demonstrates that the mechanism is not simple ion exchange.

The solution used in most stage (i) tests to sumulate high-conductivity Salem waste-water, based on analysis of plant water samples, was made up of 13% synthetic sea-water (1.365 g/L Na, 0.176 g/L Mg, 0.052 g/L Ca, 0.049 g/L K, 2.470 g/L Cl and 0.345 g/L SO_4) as well as 2.290 g/L H_3BO_3. The pH was adjusted to 7.7. The solution was doped with 5 mg/L Cok introduced as $Co(NO_3)_2 \cdot 6H_2O$. The tests were run at room temperature in columns with a height-to-diameter ratio of 40 and a flow rate of 0.6 column volume (CV) per minute. The DF10 throughputs, i.e. the numbers of column volumes (CV) of test solution passed through the column by the time that the decontamination factor has fallen to 10, were 37 CV in the case of a conventional cation exchange resin (Dowex® HCR-S) and 155 CV in the case of Durasil™ 70. The initial decontamination factor was approximately 40 at a residence time of 1.6 minute.

Stage (ii)

Stage (ii) consisted of 47 column runs using the simulated stream described above except for the use of 0.6 μCi/L Co-60 (equivalent to 0.5×10^{-3} μg/L Co) as a dopant instead of 5 mg/L of non-radioactive cobalt. Tests were run at room temperature using beds with a height-to-diameter ratio of 3.6 and flow rates of 0.2 to 4.0 column volumes per minute. The effective throughputs obtained with this test solution ath the slow end of this range of flow rates were less than 100 CV in the case of the Dowex® HCR-S resin and greater than 13000 CV in the Case of Durasil™ 70. The term "throughput", expressed in terms of the number of column volumes passed through a bed by the time it becomes exhausted, is used here in preference to the term "capacity" which is related to the total number of sorption or ion exchange sites. In the case of a selective medium, throughputs with respect to effective removal of particular species are more useful. Throughputs of non-selective resins for cobalt removal primarily depends on cobalt concentration because of its high selectivity for cobalt in the presence of an excess of foreign ions, as is always the case in nuclear plant waste-water. However, this selectivity does not amount to complete insensivity to the presence of other solutes. At the low concentrations of cobalt observed in such water the throughput is ultimately by saturation, clogging or degradation of the sorbing sites by the other components of the stream. The initial decontamination factor in the case of Durasil™ 70 primarily depends on cobalt concentration and reaches extremely high values at the low cobalt levels normally expected in reactor waste-water.

State (iii)

Stage (iii) consisted of 70 column runs on 11 different samples of waste-water taken from aqueous waste tanks at the Salem Nuclear Generating Station over a period of 3 years. Typical results obtained with the actual waste water sample which served as the model for the simulated solution used in Stage (i) and (ii) without the addition of further cobalt indicated a capacity of 80 CV with either Dowex® HCR-S cation-exchange resinor a 1:1 mix of cation-exchange and anion-exchange organic resins used in the plant, and a throughput of

1600 CV with Durasil™ 70. The residence time was 15 minutes. Results obtained monitoring Co-58 in the waste water were identical with those observed for Co-60. The concentrations of Co-60 and Co-58 in the waste water were 0.040 μCi/L and 0.134 μCi/L, respectively. The observation that Durasil™ 70 throughput is somewhat smaller than the result obtained in stage (ii) can be attributed to a higher cobalt concentration of approximately 0.1 mg/L in the actual waste water and to possible fouling by organic contaminants. The latter is always possible in nuclear plant waste-water, which frequently contains quantities of oil, soaps and other organics.[3] Subsequent testing in actual plant environments has shown Durasil™ to be somewhat less susceptible to fouling than conventional resins. However, the performance of Durasil ™ 70 is still adversely affected by the strongly chelating ingredients of decontamination solutions such as EDTA.

Stage (iv)

Tests using a column size of 1.2 L were carried out at the Salem plant itself. The test solution consisted of a 200-gallon tank of waste water similar in composition to the sample used in stage (iii) and filtered through a 5-micron filter. One test column was loaded with 1.2 L of Durasil™ 70. Another was loaded with the medium used at the time in the plant, i.e. a bottom layer of 1.05 L of mixed-bed cation-anion resin and a top layer of 0.15 L of cation resin alone. The direction of flow is from top to bottom. The results of the test showed that the throughput of the organic resin column was 30 CV, while the throughput of the Durasil™ 70 column was more than 530 CV, at which point the influent tank became empty. The average decontamination factor obtained for the organic resin column prior to loss of capacity was 10 compared with 100 in the case of the Durasil™ 70 column. Based on the results of the previous stages, the Durasil™ 70 column was operated at a residence time of 10 minutes, while a residence time of 45 minutes was allowed for the organic resin column. It was concluded that, although the Durasil™ 70 column was operated at a much faster processing rate, its throughput exceeded that of the organic resin by more than one order of magnitude and it had much higher decontamination factor. These results confirmed the observations made in the laboratory-scale studies described above.

Stage (v)

In March 1985, Durasil™ 70 was introduced into full-scale operational use at the Salem Nuclear Generating Station instead of organic resins. Both media were used in a pressurized ion exchange system introduced into use 6 months earlier. Accordingly, the performance of the organic resin bed used during this period can be directly compared with the results of the first year of full-scale operations with Durasil™ 70. The principal radio-isotopes in the Salem waste-water influent were Co-58 and Co-60, both present at levels of 0.1-1 μCi/L. The average conductivity of the waste water influent during the earlier period of the use of organic resins was approximately 790 μmho·cm^{-1} and that during the later period when Durasil™ 70 was used, 1540 mho·cm^{-1}. The throughput achieved with the organic resins was 360 CV. This value is larger than the capacities of organic resins obtained in the previous stages, where the water used had a higher conductivity of approximately 7000 μmho·cm^{-1}, corresponding to the highest range of conductivity occasionally observed in the Salem aqueous wastes. The throughput achieved with Durasil™ 70 was 5270 CV, i.e. it exceeded the throughput of the organic resins by a factor of 14, even though, as documented above, the conductivity of the water processed with Durasil™ 70 was higher by a factor of almost 2. Furthermore, the overall decontamination factor obtained during operations with Durasil™ 70 was 43, compared with a factor of 7 obtained with the organic resins. It should also be noted that due to the much larger throughput of Durasil™ 70 and the resulting longer operational lifetime of the bed, which exceeds 1 year, at least 80% of the 71-day Co-58 isotope sorbed on the column decays prior to bed exhaustion and removal, thus reducing the effective radiation load to be considered with respect to column change operations, transportation and burial. The results obtained for the performance of Durasil™ 70 for cobalt removal at Salem, compared with results obtained with mixed-bed organic resins, at each of the stages detailed above, are given in Table 1.

Table 1

Removal of Cobalt from Salem Streams

Stage	Throughput, column volumes	
	Mixed-bed organics	Durasil™ 70
i	37	155
ii	100	13000
iii	80	1600
iv	30	530
v	360	5270

ARKANSAS NUCLEAR ONE

The Arkansas Nuclear One (ANO) Generating Station in Russellville, Arkansas, is a two-unit pressurized water reactor plant operated by the Arkansas Power and Light Company. Since service water is obtained from a nearby fresh water lake, the conductivity of aqueous streams generated at ANO is generally lower to a significant extent than the conductivity of waste water at Salem (see above), although, as detailed below, at ANO high-conductivity wastes are generated and require treatment in special cases. The average conductivity of ANO waste water is (50 ± 20) μmho·cm^{-1}. The pH is 6.0-6.5.

Unlike the case of Salem, where cobalt radio-isotopes account for a predominant fraction of the total radioactivity, ANO streams usually contain significant amounts of Cs-137, Cs-134 and I-131 as well as Co-58 and Co-60. Accordingly, efforts to reduce the volume of spent ion exchange media used to decontaminate ANO streams have required the development of combinations of media which are selective towards cesium and iodide as well as cobalt. This simulated solution was doped with 5 mg/L Ca, introduced as $CaNO_3$, 5 mg/L I, introduced as NaI, and 5 mg/L Co, introduced as $Co(NO_3)3 \cdot 6H_2O$.

Stage (i)

Analysis of samples of various ANO streams has shown that the floor drains constitute the waste water component with the highest conductivity. As in the case of Salem water (see above) the composition of this component was used in formulating the simulated solution for laboratory tests of ion exchange and sorption media. This solution contained 388 mg/L Na, 5.6 mg/L K, 8.1 mg/L Ca, 1.0 mg/L Mg, 604 mg/L Cl and 2.4 mg/L SiO_2, as well as 1040 mg/L of H_3BO_3. The pH was 7.5. This simulated solution was doped with 5 mg/L Ca, introduced as $CaNO_3$, 5 mg/L I, introduced as NaI, and 5 mg/L Co, introduced as $Co(NO_3)_2 \cdot 6H_2O$.

A combination of selective media developed to remove I, Ca and Co from this solution consisted of three components. The first was Durasil™ 60, a charcoal-based anion exchanger, with a high specificity for iodine species in solution even in the presence of a large excess of anions such as chloride, sulfate, nitrate and nitrite.[1] The second element of the combination was Durasil™ 10, a high-silica porous glass cation exchanger with high selectivity with respect to cesium in the presence of high concentrations of cations such as Na^+, K^+, Mg^{2+} and Ca^{2+}.[1,4,5] The third component was Durasil™ 70, the cobalt-selective medium described above. The throughputs obtained for Cs, I and Co with the Durasil™ media were much higher than those observed in the case of the organic resins.

Stage (ii)

This stage consisted of 56 column runs with the simulated ANO solution described above doped with approximately 0.4 μCi/L Cs-137, 1.0 μCi/L I-131, and 0.6 μCi/L Co-60 instead of parts-per-million levels of Cs, I and Co. Since by the time these experiments have been conducted the improved performance of the selective media compared with conventional organic cation- and anion-exchangers were already known, these experiments mainly served to determine the relative volumes and preferred order of the three components in the column. The preferred arrangement was found to be a sequence of equal volumes of Durasil™ 60, Durasil™ 10 and Durasil ™ 70, respectively. With respect to iodide removal for instance, this combination had a throughput of 2100 CV compared with 135 CV in the case of a combination of Dowex® HCR-S and SBR cation- and anion-exchange organic resins.

Stage (iii)

Stage (iii) consisted of 10 laboratory scale column runs aimed at comparing the performance of the triple combination of selective resins optimized in stage (ii) with that of a double combination of organic cation-exchange and anion-exchange resins. The test solutions were actual ANO floor drain samples similar in composition to the simulated solution used in the previous stage. The throughputs of the 1:1:1 Durasil™ 60 - Durasil™ 10 - Durasil™ 70 column sequence for Cs-137, I-131 and Co-58, were 750 CV, >940 CV and 800 CV, respectively. (Capacities are given with respect to the total volume of the train, i.e. the sum of the volumes of the three Durasil™ media.) The corresponding throughputs of mixed-bed organic resins were 170 CV, 530 CV and 240 CV, respectively. The maximum decontamination factor obtained with the Durasil™ combination was >100 for all three isotopes, while in the case of the organic resins it was >100 for I-131 and 70 for Cs-137 and Co-58.

Stage (iv)

The laboratory tests were followed by a three-week intermediate scale on-site, on-stream test, using a combination of ANO floor drain and miscellaneous waste water. This water was passed through a 5 micron filter and then divided into two streams, one of which was passed through a sequence of column of Durasil™ 60, Durasil™ 10 and Durasil™ 70, and the other through a combination of two columns of mixed-bed organic resins. The size of each column was 0.13 L. The residence time used for the entire Durasil™ train was 2.1 minutes, and the time used for the organic combination was 4.3 minutes. The results showed the throughputs of the Durasil™ train with respect to Cs, I-131, Co-58 and Co-60 to be 1260 CV, >1900 CV, >1900 CV and 200 CV, respectively. (In the case of Cs, results for Cs-137 and for Cs-134 were identical.) The corresponding throughputs obtained with the orgainic train were 260 CV, 120 CV, 455 CV, and 70 CV, respectively. The decontamination factors obtained during the operational period for the two trains were similar in the cases of Cs (>100) and of Co (approximately 5.5, probably due to the presence of colloidal particles), while in the case of I-131 the decontamination factor was much higher in the case of the Durasil™ media (160) than in the case of the organic resins (10).

Stage (v)

Full scale processing at Arkansas Nuclear One with Durasil™ media started on September 24, 1985. During the first 14 weeks a sequence of 400 L of Durasil™ 60 and 800 L of Durasil™ 10. (Durasil™ 70 was not used during this initial period since the average levels of Co-58 and Co-60, 0.3 and 0.05 μCi/L, respectively, were low compared with the average levels of Cs-137, Cs-134 and I-131, which were 7.7, 4.8 and 0.75 μCi/L respectively.) The routine waste water combination passed through the columns during this period has a pH of 6.3 ± 0.3 and a conductivity of (50 ± 20) $\mu mho \cdot cm^{-1}$. The throughput achieved during this operational period was >2500 CV with respect to Cs-137, Cs-134, I-131 and approximately 1900 CV with respect to Co-58 and Co-60. Because of the small contribution of Co the column operations have continued after the decontamination factors with respect to Co-58 and Co-60 were observed to drop. The average decontamination factors for Cs, I and Co over the entire fourteen-week period were 1174, 217 and 45, respectively. The corresponding previous decontamination factors obtained with organic resins were 40, 3 and 10, respectively. Operations with Durasil™ media have continued in 1986, adding Durasil™ 70 to the train in cases ofincreased relative contribution of Co-58 and Co-60 to the radioactivity of the waste-water.

Table 2

Removal of Contaminants from ANO Streams

Stage	Throughput, train volumes Mixed-bed Organics			Durasil™ 60+10+70		
	Cs	I	Co	Cs	I	Co
ii	125	135	60	850	2100	500
iii	170	530	240	750	>940	800
iv	260	120	455	1260	>1900	>1900
v	550	440	440	>2500	>2500	1900

The results obtained for the performance of Durasil™ media for contaminant removal at ANO, compared with results obtained with mixed-bed organic resins, at each of the stages detailed above, are given above. Throughput values are given of the entire volume of the train of Durasil™ or organic media. Stage i results are not given since at this stage tests were run on individual media (e.g. Durasil™ 10) rather than on combinations (trains) of several beds.

SPECIAL OPERATIONS - HIGH CONDUCTIVITY WATER

In addition to ordinary waste water, Arkansas Nuclear One separately accumulated in 1984 a volume of 240 m^3 of waste-water with a conductivity of 12000 $\mu mho \cdot cm^{-1}$ and a pH of 11.68. The most important constituents were sodium borate (approximately 1.5 g/L) and sodium nitrite (approximately 0.6 g/L). The major radioactive contaminants were Cs-137, Cs-134, Co-60 and Co-58, present at levels of 25, 7.3, 2.5 and 1.2 μCi/L, respectively. Other radio-isotopes, such as Mn-54 and Sb-122, were present at much lower levels (0.02 μCi/L). Nineteen laboratory scale stage (iii) experiments were carried out on this water. They showed that total throughputs of approximately 500 CV with respect to Cs and Sb and 800 CV with respect to Co could be achieved with a sequence of Durasil™ 10 and Durasil™ 70 columns following neutralization of the solution with H_2SO_4. However, due to the high levels of radioactivity only about 30% of the total throughput was utilitzed in subsequent plant operations in order to maintain high decontamination factors. Following the laboratory tests the entire volume of ANO high-conductivity waste water was decontaminated using a combination of 0.62 m^3 of Durasil™ 10 and 0.85 m^3 of Durasil™ 70. The total throughput was accordingly 155 CV. The initial decontamination factors for Cs, Co, Sb and Mn were 1400, 900, 60 and 100, respectively, and the corresponding throughput of such resins was 17 CV and the initial decontamination factors 40 for Cs and 10 for Co.

The full-scale use of Durasil™ media to decontaminate the ANO high-conductivity water thus resulted in improvements by about one order of magnitude in both capacity and decontamination factors.

CONCLUSIONS

The use of selective ion exchange media for radioactive contaminants in nuclear plant waste water has been demonstrated to give considerable improvements in throughput, decontamination factors and processing rates compared with the use of conventional resins. At the same time, such improvement requires comprehensive analysis of the chemistry and radiochemistry of the stream to be processed, followed by careful selection and optimization of suitable specialized media. This can be accomplished through a series of steps comprising laboratory studies with non-radioactive and radioactive simulated solutions and, subsequently, with samples of actual plant water, followed by intermediate-scale on-site tests, finally leading to full-scale plant operations.

LITERATURE CITED

1. Aa. Barkatt, K.A. Michael, W. Sousanpour, Al. Barkatt, L.M. Penafiel, P.B. Macedo and H.G. Sutter, Properties of Novel Selective Ion Exchangers for Nuclear Plant Applications, Nucl. Tech., **78,** 75-82 (1987).

2. Aa. Barkatt, P.B. Macedo, L.M. Penafiel, and W. Sousanpour, Removal of Radioactive Cobalt from Highly Saline Streams, Trans. Am. Nucl. Soc., **45,** 160-162 (1983).

3. Aa. Barkatt, P.B. Macedo, L.M. Penafiel, W. Sousanpour and H.G. Sutter, Removal of Radioactive Cesium from Streams with a High Potassium Content, Trans. Am. Nucl. Soc. **46,** 167-168 (1984).

4. Aa. Barkatt, K.A. Michael, W. Sousanpour, Al. Barkatt, L.M. Penafiel, P.B. Macedo and H.G. Sutter, The Use of an Inorganic Glassy Cation Exchanger for the Purification of Nuclear Waste Streams, Proc. 46th Int. Water Conf., Engineers Soc. Western Pennsylvania, Pittsburgh, PA, 1985, pp. 140-244.

NUMERICAL SIMULATION OF FIXED-BED ADSORPTION DYNAMICS BY THE METHOD OF LINES

Ben F. Brian III and Imre Zwiebel ■ Department of Chemical, Bio and Materials Engineering, Arizona State University, Tempe, AZ 85287

Ronald S. Artigue ■ Department of Chemical Engineering, Rose-Hulman Institute of Technology, Terre Haute, IN 47803

The method of lines (MOL) with finite difference discretization is demonstrated as an alternative to traditional orthogonal collocation and finite difference methods for the solution of an isothermal model, accounting for both axial and intraparticle dispersion, of fixed-bed adsorption dynamics. In this unique usage of the MOL the second order differentials are approximated by central differencing while backward differencing is used for the first order terms. The MOL algorithm is robust and allows specification of the finite (rather than zero) concentration gradient boundary condition at the exit of the bed. Breakthrough curves predicted by the MOL are compared to published (Raghavan & Ruthven, *AIChE J.*, 29, 922, 1983) semi-analytical and orthogonal collocation solutions of a simplified single adsorbate system. The agreement is excellent over a wide range of parameters.

INTRODUCTION

The development and numerical solution of mathematical models which describe the dynamics of fixed bed adsorption have received considerable attention in the literature. Standard finite difference and orthogonal collocation methods have been used in the majority of the studies. In this paper we present the method of lines (Liskovets, 1965; Davis, 1984) with finite difference discretization as an alternative to these techniques. For the numerical solution of partial differential equations in time and spatial dimensions, the method of lines involves discretization of the spatial gradients only, and thus the PDE system is approximated by a coupled system of ODE's in the time domain. It is the discretization method that distinguishes the various techniques. Actually, the method of orthogonal collocation is a form of the method of lines with collocation chosen as the discretization method. Various finite difference discretization techniques for the method of lines are discussed by Carver (1976).

The method of lines (MOL) has previously been applied to the simulation of fixed bed adsorption dynamics by Wang and Niedzwiecki (1982), who used forward differencing discretization to solve a multicomponent adiabatic model. They neglected axial and intraparticle dispersion, thus resulting in a hyperbolic system of equations. Costa and Rodigues (1985) applied the method of lines with orthogonal collocation discretization to a model with both axial and intraparticle diffusion. In this paper, we present a unique finite difference MOL solution of a fixed bed adsorption model accounting for both axial and intraparticle dispersion. To illustrate the application of the MOL, we have solved the case of a single adsorbate model with linear equilibrium, for which semi-analytical and orthogonal collocation solutions have been reported (Raghavan and Ruthven, 1983). A similar model with nonlinear equilibrium has been solved using the MOL (Brian, 1985), and the increased complexity was found to be of no consequence. The availability of semi-analytical and collocation solutions for the simplified model provides a check on the accuracy of the MOL solutions.

MATHEMATICAL MODEL

The mathematical model presented in this study is identical to the problem solved by Raghavan and Ruthven (1983). The modeled system is an isothermal adsorption column packed with spherical adsorbent particles. At time zero, the column is subjected to a small step change in the concentration of an adsorbable species in the carrier fluid. The flow external to the particles is described by a dispersed plug flow model. Intraparticle diffusion resitance is included and instantaneous linear equilibrium is assumed inside the particles. The system is described by the following mass balance equations:

Particle: $$\frac{\partial q}{\partial t} = D\left(\frac{\partial^2 q}{\partial r^2} + \frac{2}{r}\frac{\partial q}{\partial r}\right) \qquad (1)$$

External Fluid: $$\frac{\partial c}{\partial t} = D_L \frac{\partial^2 c}{\partial z^2} - v\frac{\partial c}{\partial z} - \left(\frac{1-\varepsilon}{\varepsilon}\right)\left\{\frac{3k}{R}\left(c(z,t) - \frac{q|_{r=R}}{K}\right)\right\} \qquad (2)$$

Initial and Boundary Conditions: $$q\,(r,t=0) = 0 \qquad (3)$$

$$\frac{\partial q}{\partial r}\Big|_{r=0} = 0 \qquad (4)$$

$$D\frac{\partial q}{\partial r}\Big|_{r=R} = k\left[c(z,t) - \frac{q|_{r=R}}{K}\right] \qquad (5)$$

$$c(z,t=0) = 0 \qquad (6)$$

$$D_L \frac{\partial c}{\partial z}\Big|_{z=0} = -v\,(c|_{z=0} - c|_{z=0+}) \qquad (7)$$

$$\frac{\partial c}{\partial z}\Big|_{z=L} = 0 \qquad (8)$$

There has been considerable discussion of the validity of the boundary conditions used at the column exit in dispersed flow systems (Bischoff, 1961; Fan and Ahn, 1962). Equation (8) has generally been the accepted form. However, our finite difference MOL solution allows an explicit approximation of Equation (2) at the exit of the bed, resulting in the boundary condition:

$$\frac{\partial c}{\partial z}\Big|_{z=L} = \text{finite} \qquad (9)$$

Incorporating the dimensionless variables (Q, U, τ, x, η, ξ, Pe, Ψ, Θ) used by Raghavan and Ruthven (1983), Equations (1) to (9) become:

$$\frac{\partial Q}{\partial \tau} = \nabla^2 Q \qquad (10)$$

$$\frac{\partial U}{\partial \tau} = \frac{1}{Pe}\Psi\Theta\frac{\partial^2 U}{\partial x^2} - \Psi\Theta\frac{\partial U}{\partial x} - 3\Psi\xi\left(U - \frac{Q|_{\eta=1}}{K}\right) \qquad (11)$$

$$Q(\eta,\ \tau=0) = 0 \qquad (12)$$

$$\frac{\partial Q}{\partial \eta}\Big|_{\eta=0} = 0 \qquad (13)$$

$$\frac{1}{K}\frac{\partial Q}{\partial \eta}\Big|_{\eta=1} = \xi\left(U - \frac{Q|_{\eta=1}}{K}\right) \qquad (14)$$

$$U\,(x,\ \tau=0) = 0 \qquad (15)$$

$$\frac{\partial U}{\partial x}\Big|_{x=0} = -Pe\,(U|_{x=0} - U|_{x=0+}) \qquad (16)$$

$$\frac{\partial U}{\partial x}\Big|_{x=1} = 0 \qquad (17)$$

$$\frac{\partial U}{\partial x}\Big|_{x=1} = \text{finite} \qquad (18)$$

Equations (10) and (11) (with the appropriate boundary and initial conditions) are the set of coupled linear parabolic PDE's that were solved using the method of lines.

NUMERICAL SOLUTION BY THE METHOD OF LINES

In our method the adsorption column length and particle radius are discretized into NB and NR points, respectively. The spatial derivatives in Equations (10) and (11) are then approximated using finite differences. This transforms these equations into NB + NB(NR) ODE's in τ. Three point central differencing is used to approximate the gradients within the particle (Equation 10), and we have found the PVW (Price, Varga and Warren, 1966) finite difference analogs to be best suited for the approximation of the axial gradients in Equation (11). The PVW formulas use central differencing for the 2nd order term and backward differencing for the 1st order term (both are second order correct), and are given (for axial discretization location j) by:

$$\left.\frac{\partial^2 U}{\partial x^2}\right|_j = \frac{U_{j+1} - 2U_j + U_{j-1}}{\Delta x^2} \tag{19}$$

$$\left.\frac{\partial U}{\partial x}\right|_j = \frac{3U_j - 4U_{j-1} + U_{j-2}}{2\Delta x} \tag{20}$$

Therefore when Equation (11) is written at axial location j, Equations (19) and (20) are substituted and it becomes:

$$\frac{\partial U_j}{\partial \tau} = \frac{1}{Pe}\,\Psi\Theta\,\frac{U_{j+1} - 2U_j + U_{j-1}}{\Delta x^2} - \Psi\Theta\,\frac{3U_j - 4U_{j-1} + U_{j-2}}{2\Delta x} - 3\Psi\xi\,\left(U_j - \left.\frac{Q|_{\eta=3}}{K}\right|_j\right) \tag{21}$$

At the column exit, or j=NB, a dummy point outside the column is used to evaluate the boundary condition given by Equation (17). When Equation (18) is used as the boundary condition, a higher order backward difference formula is used to approximate the second order term in place of Equation (19). In this manner, at the bed exit all the spatial gradients are written in terms of the concentrations at axial positions within the column, and no finite difference approximation of a boundary condition is required. The initial value problem is then solved using the International Math and Statistics Library (IMSL) version of Gear's method (Hindmarsh, 1974). The solution method is presented in detail elsewhere (Brian, 1985).

RESULTS AND DISCUSSION

Stability Analysis

In a similar fashion to Costa and Rodrigues (1985), a series of simulations were carried out with various Peclet numbers and different numbers of axial discretization points (NB) to determine the number of points required for a given accuracy of solution. The dimensionless parameters were chosen so that the film and intraparticle resistances were as small as possible. In this manner, the profiles would be sharper and our tests more conservative. Three factors were compared in these studies: (i) the maximum absolute error between the solutions (for a given value of NB), and the final converged solutions (NB=60); (ii) the magnitude of any normalized concentration below zero (there were no values greater than 1.0); (iii) CPU time on an IBM 3081 system. The results of these tests for boundary conditions Equation (17) and (18) are shown in Tables 1 and 2, respectively. From these results it is evident that at high Peclet numbers a larger number of discretization points are required. This is consistent with collocation solutions (Costa and Rodrigues, 1985), and stems from forcing a second boundary condition on a system of equations that is increasingly hyperbolic. At low Peclet numbers the MOL solutions converge for a small number of discretization points. At a set value of NB, there is no significant increase in CPU time with decreasing Peclet number. This was not true for collocation solutions (Raghavan and Ruthven, 1983). Finally, by comparing Tables 1 and 2 it is evident that neither boundary condition is superior in terms of stability.

Effect of the Exit Column Boundary Condition

Figure 1 shows the effect of the exit column boundary condition on the calculated breakthrough curves for large and small Peclet numbers. For large Peclet numbers the curves are identical while at small values of the Peclet number large differences occur. The finite difference approximation to Equation (17) results in the stipulation that the concentration one step size length beyond the column exit is equal to the concentration one step size length inside the column, and thus requires continuity of the column geometry across the boundary. On the other hand, with Equation (18) the concentration at the column exit is solved in terms of the concentrations inside the column, and no assumptions are made across the boundary. However, we have found that the boundary condition given by Equation (18) is the finite difference equivalent to the infinite boundary condition given by

$$U\ (x=\infty,\ t) = 0 \tag{22}$$

It is not our goal to argue the validity of either boundary condition, but in terms of finite difference approximations the condition given by Equation (22) or (18) is physically more realistic than Equation (17). Our method provides a convenient finite difference solution with an infinite boundary condition.

Comparison of Our MOL Solutions to Published Results

Raghavan and Ruthven (1983) compared their orthogonal collocation solutions with semi-analytical solutions provided by Rasmuson and Neretnieks (1980) for the model described by Equations (10) to (17). As a check on the accuracy of our method, points were taken from their comparisons and are shown in Figures 2 and 3 along with our MOL solutions. From these figures it is evident that the method of lines solutions are in excellent agreement with the other methods for the selected values of the parameters.

In conclusion, we have developed a finite difference method of lines solution to a complicated mathematical model of fixed bed adsorption. The solution method has excellent convergence properties at low Peclet numbers and is consistent with the convergence of other techniques at high Peclet numbers. Our solutions are in excellent agreement with published semi-analytical and orthogonal collocation solutions. A unique feature of our solution method is that it allows specification of a finite (rather than zero) gradient boundary condition at the column exit, which results in the finite difference equivalent of the infinite fixed column adsorber boundary condition.

NOTATION

c = fluid phase-concentration, mol/m^3

c_o = fluid-phase inlet concentration, mol/m^3

D = solid-phase diffusivity, m^2/s

D_L = axial dispersion coefficient, m^2/s

j = dimensionless axial position index

K = volume equilibrium constant, m^3/m^3

k = mass transfer coefficient, m/s

L = length of adsorber, m

NB = number of discretization points in the external fluid phase

NR = number of discretization points in the particle (internal) phase

Pe = vL/D_L, Peclet number

Q = q/c_o, dimensionless internal concentration in particles

q = internal concentration in particles, mol/m^3

R = particle radius, m

r = radial distance from center of spherical particle, m

t = time, s

U = c/c_o, dimensionless fluid-phase concentration

v = average interstitial velocity, m/s

x = z/L, dimensionless axial distance

z = axial distance coordinate, m

Greek Letters

Δ = dimensionless distance between two discretization points

Θ = $vR^2\varepsilon/LDK\ (1 - \varepsilon)$, bed length parameter

ε = bed porosity, m^3/m^3

η = r/R, dimensionless radial distance in particle

∇^2 = Laplacian operator

ξ = kR/DK, film resistance parameter

Ψ = $K((1 - \varepsilon)/\varepsilon)$, distribution ratio

τ = Dt/R^2, contact time paramter

LITERATURE CITED

Bischoff, K.B., "A Note on Boundary Conditions for Flow Reactors," Chem. Eng. Sci., 16, 131 (1961).

Brian, B.F., "Numerical Simulation of Fixed Bed Adsorption Dynamics by the Method of Lines," MS thesis, Arizona State University, (1985).

Carver, M.B., "The Choice of Algorithms in Automated Method of Lines Solution of Partial Differential Equations," Numerical Methods for Differential Systems (Ed. Lapidus, L. and W. E. Schiesser), Academic Press (1976).

Costa, C., and A. Rodrigues, "Design of Cyclic Fixed-Bed Adsorption Processes Part 1: Phenol Adsorption on Polymeric Adsorbents," AIChE J., 31, 1645 (1985).

Davis, M.E., Numerical Methods and Modeling for Chemical Engineers, John Wiley and Sons (1984).

Fan, L., and Y. Ahn,, "Critical Evaluation of Boundary Conditions for Tubular Flow Reactors," I&EC Proc. Des. and Dev., 1, 190 (1962).

Hindmarsh, A.C., "GEAR: Ordinary Differential Equation System Solver," Lawrence Livermore Laboratory, Report UCID-30001, Revision 3, December (1974).

Liskovets, O.A., "The Method of Lines," J. Diff. Eqns., 1, 1308 (1965).

Price, H.S., R.S. Varga, and J.E. Warren, "Application of Oscillation Matrices to Diffusion - Convection Equations," J. Math and Physics, 45 301 (1966).

Raghavan, N.S., and D.M. Ruthven, "Numerical Simulation of a Fixed-Bed Adsorption Column by the Method of Orthogonal Collocation," AIChE J., 29, 922 (1983).

Rasmuson, A. and I. Neretnieks, "Exact Solution of a Model for Diffusion in Particles and Longitudinal Dispersion in Packed Beds," AIChE J., 26, 686 (1980).

Wang, Y. W., and J. L. Niedzwiecki, "A Simplified Model for Multicomponent Fixed Bed Adsorption," AIChE Symp. Series, 219, 120 (1982).

Table 1. Precision and stability of the method of lines solutions using the exit column boundary condition given by Equation (17).

	NB				
Pe	10	20	30	40	50
200	0.06* -.046** 3.49***	0.021 -0.01 24.96	0.009 -0.002 83.73	0.004 -- 203.39	0.002 -- 434.29
100	0.049 -0.037 3.88	0.017 -0.005 26.18	0.008 -- 92.27	0.003 -- 217.58	0.001 -- 418.03
50	0.037 -0.024 3.98	0.013 -0.001 25.47	0.006 -- 90.47	0.003 -- 224.22	0.001 -- 446.09
5	0.005 -- 3.89	0.002 -- 18.41	0.001 -- 88.19	0.001 -- 249.02	0.001 -- 540.06

* Maximum Absolute Error
** Below Zero Oscillation
*** CPU Time, NR=5 (sec)

Table 2. Precision and stability of the method of lines solutions using the exit column boundary condition given by Equation (18).

	NB				
Pe	10	20	30	40	50
200	0.060* -0.046** 3.99***	0.020 -0.010 26.95	0.008 -0.001 90.46	0.004 -- 217.63	0.002 -- 437.24
100	0.049 -0.037 3.95	0.016 -0.005 26.06	0.006 -- 90.82	0.003 -- 215.39	0.001 -- 422.63
50	0.035 -0.024 3.87	0.011 -0.001 24.71	0.004 -- 89.11	0.002 -- 223.44	0.001 -- 450.99
5	0.003 -- 3.67	0.001 -- 18.08	0.001 -- 89.12	0.001 -- 243.0	0.000 -- 553.12

* Maximum Absolute Error
** Below Zero Oscillation
*** CPU Time, NR=5 (sec)

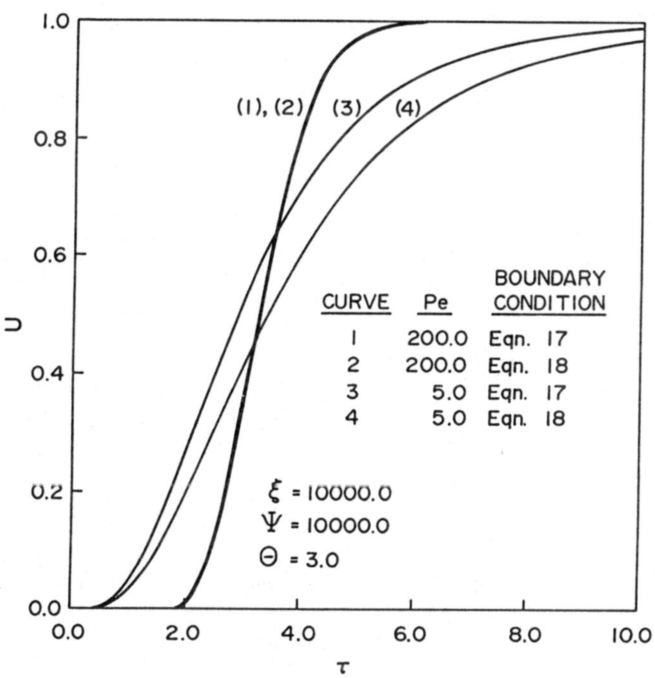

Figure 1. Breakthrough curves calculated by the method of lines showing the effect of the exit boundary condition for large and small Peclet numbers.

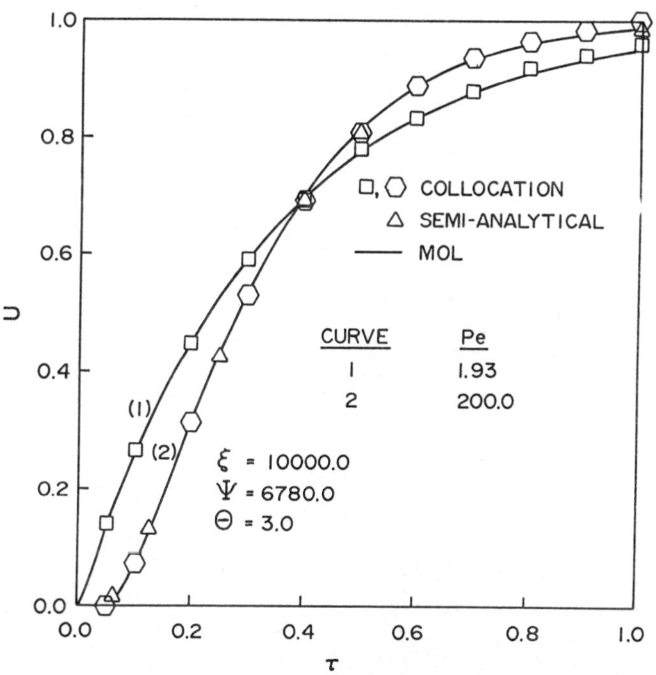

Figure 2. Comparison of the method of lines (MOL) solutions from the present study with orthogonal collocation and semi-analytical solutions of Raghavan and Ruthven (1983) [the semi-analytical solutions were originally from Rasmuson and Neretnieks (1980)].

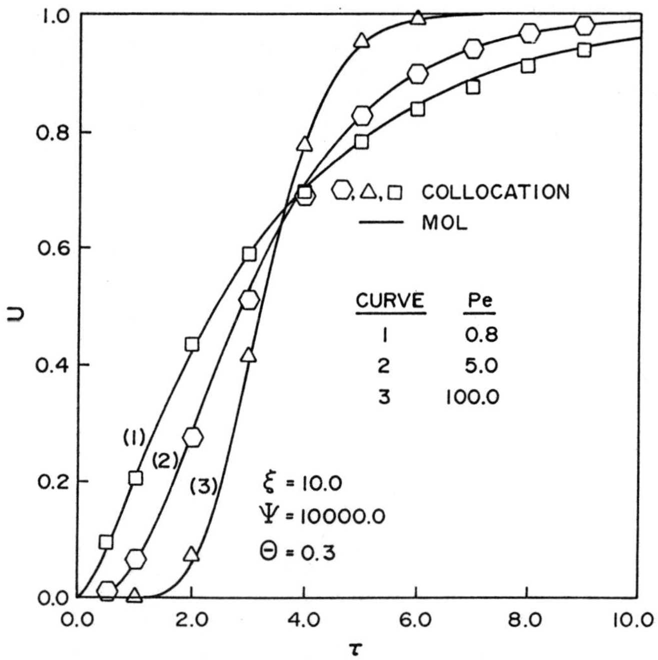

Figure 3. Comparison of the method of lines (MOL) solutions from the present study with orthogonal collocation solutions of Raghavan and Ruthven (1983) showing the effect of Peclet number.

ADSORPTION OF MIXTURES OF WATER VAPOR AND HYDROCARBONS BY ACTIVATED CARBON BEDS: THERMODYNAMIC MODEL FOR ADSORPTION EQUILIBRIUM AND ADSORBER DYNAMICS

S.J. Doong and R.T. Yang ■ Department of Chemical Engineering, State University of New York at Buffalo, Buffalo, NY 14260

In Part I, a simple thermodynamic model is developed for predicting the equilibrium adsorption of water vapor mixed with hydrocarbons on activated carbon. The model is simple to use and compares favorably with experimental data. This model is used to predict the adsorber bed behaviors during adsorption and desorption of a binary (water-benzene) mixture. During adsorption, a long leading wavefront followed by a shock wave for water vapor exists due to the S-shaped isotherm, and a significant amount of water condensation can occur.

Water vapor is usually present in industrial gas purification and solvent recovery processes, where activated carbon is the standard sorbent. However, studies on the adsorption of mixed water vapor and hydrocarbons on activated carbon rarely appeared in the literature. A unique feature of this important system is that the water/carbon isotherm is S-shaped, whereas the hydrocarbon isotherms are of the Langmuir type. This unique feature makes the predicition of the mixed gas adsorption equilibrium and hence adsorber dynamics a difficult problem.

This is a two-part paper. In the first part, a thermodynamic model is presented which can be used to predict equilibrium adsorption from mixtures of water vapor and hydrocarbons on activated carbon, based on pure-gas isotherm data. The second part addresses the dynamics of an adsorber for the same gas-solid system, using the termodynamic model given in Part I.

Part I. EQUILIBRIUM ADSORPTION OF MIXTURES OF WATER VAPOR AND HYDROCARBONS

Introduction

Activated carbon is widely used for gas purification and solvent recovery where organic vapors are adsorbed. Water vapor frequently is contained along with the organic vapor. A problem of particular interest is the adsorption of chemical agents from the ambient air. A quantitative model is needed for predicting, based on pure-gas isotherms, the adsorption from these mixtures. This practical and important problem has attracted little attention.

The adsorption of mixtures of organic vapors and non-polar or weakly polar gases can be predicted reasonably well by several theories, e.g., ideal adsorbed solution theory (IAS) (1), extended Langmuir and the loading ratio correlation (LRC) methods (2), and the potential theory (3). These methods, however, all fail when water vapor is contained in the mixture. A reason for the failure is that the water vapor/activated carbon system exhibits a peculiar behavior characterized by an S-shaped (Type V) isotherm and a well-defined hysteresis region (4). Much effort has been made on the isotherm of water vapor on activated carbon (5-7). Yet, no single isotherm equation is available that can describe the entire isotherm for this system. Prediction of mixture adsorption for water/organic vapor is obviously just more difficult.

It is the intent of this work to establish a model equation that can be used for adsorber calculations. As such, the equations must be simple and require the least possible amount of iteration in calculation.

Okazaki, et al. (8) appear to be the first to attack this problem. They presented both detailed experimental data and a model. Their model is based on evaluating the amounts of solvent and water in the capillary-condensed phase in the fine pores and, separately, in the ordinary adsorbed phase in the large pores. The necessary data for the model are: the pure-component isotherms, the liquid-phase isotherms of organics, and the vapor-liquid equilibria. In addition, the pore volume and surface area distribution curves for the carbon are required. The predicted results agree fairly well with their experimental data. Although straightforward, their calculation procedure demands a great deal of information on the thermodynamic and physical properties for the system. It would be very cumbersome to implement their method in computer simulation for fixed-bed design purposes.

More recently, Manes (9) proposed a potential-theory model for estimating the effect of humidity on the adsorption of the water-immiscible organic vapor in activated carbon. Manes assumes a two-phase adsorbate and predicts zero adsorption of water vapor whenever the volume of organic adsorbate exceeds the adsorbate volume of organic-free water vapor. It is well-suited for model simulation of fixed-bed adsorption. As an iteration procedure is required in this method, the convergence of the solution depends on the initial value. Results predicted by Manes' method are also compared in this study. It has been found that his method is not easy to converge and it requires an initial value that is close to the final solution.

In this paper, we present a potential-theory method for calculating the adsorption amounts of the organic and water vapor mixture on activated carbon. The method is based on Dubinin's pore volume filling theory. Iteration procedures are also required to find the solution. However, for the same initial guess values, this method converges more easily than does Manes' method.

Pore Volume Filling Theory for Pure Non-Polar Gases

A detailed description of this theory can be found in ref. (10,11). The original isotherm equation according to the theory is the Dubinin-Radushkevich (D-R) equation:

$$v = v_o \exp[-B(\frac{T}{\beta})^2(\ln \frac{P_s}{P})^2] \qquad (1)$$

where v is the volume adsorbed at relative pressure P/P_s and temperature v_o is the limiting volume of adsorption or the total volume of the micropores, β is the affinity coefficient of the characteristic curve, and B is the parameter determined by the carbon sorbent or characteristic energy of adsorption. As micropores account for most of the surface area of activated carbon, physical adsorption should mostly take place in the micropores rather than mesopores and macropores. Furthermore, Dubinin postulated that the adsorption on microporous adsorbents involves not a successive formation of adsorbtion layers on the surface of micropores but the filling of their adsorption space.

Equation (1) is only valid for microporous sorbents. The exponent 2 in this equation is replaced by values greater than 2 for sorbents with very fine micropores.

A modification for the D-R equation was made by Dubinin et al. (10,12), shown as follows:

$$v = v_o \exp \{-B(\frac{T}{\beta})^2[(\ln \frac{P_s}{P})^2 - (\ln h_o)^2]\} \quad (2)$$

This modified equation is the result of examination of the hysteresis loops of many non-polar gases on the same activated carbon, and provides a better understanding of the adsorption. Owing to the narrow character of the micropore volume distribution, the filling of micropores is practically completed before the beginning of capillary condensation of vapor in transitional pores. Experiments showed that in adsorption of different vapors the adsorption volume at the beginning of hysteresis corresponds to a constant value, which can be regarded as microporous volume, v_o. Therefore, it is expedient to consider this value v_o at equilibrium relative pressure h_o as the limiting volume, and the value h_o is equal to the relative pressure at the beginning point of the hysteresis loop. The pore radius corresponding to the same point may be calculated from the Kelvin equation, which is nearly a constant value of 15 Å for all gases tested (10-12). This pore radius in effect is the demarcation between micropores and transition (meso- and macro-) pores. Consequently, the value for h_o can be determined from the pore structure independent of the adsorption experiment.

The success of the D-R and modified D-R equations is limited to non-polar gases. A modification to the D-R equation was made by Razwadowski et al. (13,14) for polar vapors. The reference state P_s in the potential theory is modified by multiplying an activity coefficient ν to account for the different degrees of association in the liquid and the adsorbed phases. As a result, the saturated vapor pressure of a polar gas at a given stage of adsorption, P_s^*, is the product of saturated vapor pressure P_s and an activity coefficient:

$$P_s^* = \nu \cdot P_s \quad (3)$$

The deviation of from unity actually indicates the interactions between sorbates as the free energy for the interactions is $-RT \ln \nu$ (13,14). The value of ν is obtained empirically and is dependent on P/P_s. For the adsorption of alcohols and amines, the values of ν differ substantially from unity at small amount adsorbed and approach unity as the amount adsorbed is increased.

The two modified D-R equations, Eqs. (2) and (3), will be used below for predicting mixture adsorption. The values for ν for water vapor at low values of amount adsorbed will be calculated with the aid of another isotherm equation developed recently by Dubinin, and $\nu = 1$ at higher amounts adsorbed. The addition isotherm equation was developed based on the pore-filling theory, by using the concept of dynamic equilibrium (5,6):

$$(v + \bar{v})(1 - k_1 v) \frac{P}{P_s} = k_2 v \quad (4)$$

This is a three-constant isotherm ($\bar{v}$, k_1 and k_2), which fits experimental data of water vapor/activated carbon well up to $P/P_s = 0.6$ (6,7).

Working Model for Adsorption of Water-Organic Mixtures on Activated Carbon

Based on the modified D-R equations, the following combined equation is obtained:

$$v = v_o \exp -\{(\frac{1}{\beta E})^2[(\ln \frac{\nu P_s}{P})^2 - (\ln h_o)^2]\} \quad (5)$$

As mentioned, v_o is the total pore volume which can be measured independently. E is the energy constant characteristic of the specific carbon sorbent. Both E and v_o can be determined with a standard reference vapor, namely benzene. The values of ν are assumed unity for organic vapors that are not highly polar, and are calculated for water vapor using equation (4) for $P/P_s < 0.6$ while it is assumed unity above this relative pressure. Thus, for both water vapor and the organic vapors, there are only two fitting constants (β and h_o) to be determined from pure-gas isotherm data. The assumption of $\nu = 1$ for organic vapors is justified for the four compounds studied here, as linear plots are obtained for $\ln v$ vs. $[\ln(P_s/P)]^2$ from pure-vapor data.

Equation (5) may be extended to mixture adsorption. Using the binary mixture as an example, the adsorbed volume for component 1 in the presence of component 2 is:

$$v_1 = (v_o - v_2)\exp\{-(\frac{1}{\beta_1 E})^2[(\ln \frac{\nu_1 P_{s1}}{P_1})^2 - (\ln h'_{o1})^2]\} \quad (6)$$

The available total pore volume becomes $v_o - v_2$. The value of h'_{o1} in the mixture is also different from that for its pure-component value, as less micropore volume is now available. The value of h'_{o1} should be lower than h_1, and should be a function of the composition of the mixture. In order to calculate the values for h'_{o1}, the following simple linear interpolation rule is used.

The values of h_o for pure-component and mixed-component are related by:

$$K \ln h'_{o1} = \ln h_{o1} \quad (K \leq 1) \quad (7)$$

The following are two limiting values for K:

$$K = 1 \quad \text{for} \quad v_2 = 0$$

$$\text{and} \quad K = 0 \quad \text{for} \quad v_2 = v_o \quad (8)$$

The following linear interpolation rule is used:

$$K = 1 - \frac{P_2^o}{\nu_2^o P_{s2}} \quad (9)$$

where P_2^o is the pressure calculated from pure-component isotherm for component 2 at the amount adsorbed v_2, which is the amount adsorbed in the mixture. P_2^o is a function of v_2. ν_2^o is the activity coefficient corresponding to P_2^o.

Using the linear interpolation, the following equations are for a binary mixture of water vapor and organic vapors:

$$v_1 = (v_o - v_2)\exp\{-(\frac{1}{\beta_1 E})^2[(\ln \frac{\nu_1 P_{s1}}{P_1})^2 - (\frac{\ln h_{o1}}{1 - \frac{P_2^o}{\nu_2^o P_{s2}}})^2]\} \quad (10a)$$

$$v_2 = (v_o - v_1)\exp\{-(\frac{1}{\beta_2 E})^2[(\ln \frac{\nu_2 P_{s2}}{P_2})^2 -$$

$$\left(\frac{\ell n\, h_{02}}{o}\right)^2]\} \qquad (10b)$$
$$1 - \frac{P_1}{\nu_1^o P_{s1}}$$

Equations (10a) and 10b) are nonlinear and coupled. They may be solved by iteration methods. The required pure-gas data are β_1, β_2, h_{o1} and h_{o2}. The sorbent constants v_o and E_o may be either calculated from either one of the pure-gas data or measured independently. Similar equations may be written for multicomponent mixtures.

Equations (10a) and 10b) have been used successfully to predict binary mixture data for polar and non-polar adsorptive gases on activated carbon and zeolites. This paper will only include binary mixtures of water vapor and organic vapor on carbon.

Results and Discussion

The only experimental data in the literature on adsorption of binary mixtures of water vapor and organic vapors on activated carbon are those given in Ref. (8). Two types of activated carbon with different pore-volume distributions were used. For each carbon, four binary mixtures were studied: two with water-soluble organic solvents (methanol and acetone) and two with water-insoluble solvents (benzene and toluene).

Equation (5) is first fitted with their pure-gas isotherms. The value of β for benzene is unity since benzene is used as the reference vapor. However, h_o cannot be calculated since no hysteresis data were given. Thus, h_o is a fitting parameter. However, the values of v_o are given for both carbons. The E_o values for both carbons are calculated based on benzene isotherms. From the pure-gas data, the constants for Eq. (5) are given in Table I.

In table 1, the low β value for water vapor reflects the hydrophobic nature of carbon surface, and that $\ell n\, h_o = 0$ $(P/P_s = 1)$ is also consistent to that fact since water vapor is barely adsorbed in mesopores and macropores before reaching saturation. The HGI carbon has a wider pore volume distribution than the S carbon. A large fraction of the pore volume in the carbon is in the transition pore range (pores larger than 16 Å radius). However, the proposed model works equally well for both carbons.

The comparisons between all available experimental data and model predictions for binary mixtures of benzene, acetone, methanol and toluene with water vapor adsorbed on Shirasagi S carbon and HGI carbon are listed in Tables 2-7.
Predictions by two other models are also compared in these tables. They are the IAS theory (1) and the Manes model (9). The pure-gas isotherms used for the IAS calculations are Eq. (5) and data in Table 1. Thus, the comparisons are made on the same basis.

The comparisons with experimental data show that the proposed model, Eqs. (10a) and (10b), and the Manes model are both fairly good, while the IAS fails badly as predicted. The model proposed here is, however, the simplest to use among the three. It requires no adjusting parameters.

Part II. Fixed-bed Adsorption of Mixture of Benzene and Water Vapor on Activated Carbon

Introduction

The behavior of a fixed-bed adsorber for multicomponent mixture systems has been intensively studied. The pioneering work using the method of characteristics and wave theory has unveiled a great number of properties and special features of such systems (15-19).

This work addresses the particular system of adsorption of water vapor and hydrocarbon on activated carbon. Benzene is used as the hydrocarbon, which exhibits favorable isotherms, whereas water vapor exhibits unfavorable behavior in large portions of the isotherms. Both the method of characteristics and direct numerical integration method will be used to solve for adsorption and desorption. Only preliminary results are presented here as detailed and more extensive results are presented elsewhere.

Mass Balance Equation

The system under consideration is dilute benzene vapor present in a humid carrier gas, e.g., air. The adsorbates are benzene and water; the carrier gas is assumed to be inert. The adsorbent is the Type S carbon referenced in Part I. For isothermal operation, the mass balance for each adsorbate is:

$$\varepsilon \frac{\partial C_i}{\partial t} + u \frac{\partial C_i}{\partial z} - S_i = 0 \qquad (1)$$

where ε = void fraction of the adsorber
C_i = gas phase concentration, mole/cm^3.
u = linear superficial velocity of the carrier gas, cm/sec.
S_i = local sorption rate, mole/s/cm^3.

Under equilibrium conditions, S_i is replaced by $-\rho_B \frac{\partial q_i}{\partial t}$. With the assumption of ideal gas law, we obtain

$$\varepsilon \frac{\partial P_i}{\partial t} + u \frac{\partial P_i}{\partial z} + RT\, \rho_B \frac{\partial q_i}{\partial t} = 0 \qquad (2)$$

where ρ_B is the bulk density of the bed (g/cm^3), q_i is adsorbed phase concentration (mole/g.carbon), P_i is the partial pressure of component i (mmHg). For a binary system, Equation (2) consists of two partial differential equations, which can be solved by either a numerical finite difference method or by the method of characteristics. For the direct numerical method, Equation (2) can be rearranged into:

$$\left(\varepsilon + RT\rho_B \frac{\partial q_i}{\partial P_i} \right) \frac{\partial P_i}{\partial t} + RT\, \rho_B \sum_{i \neq j} \frac{\partial q_i}{\partial P_j} \frac{\partial P_j}{\partial t} + u \frac{\partial P_i}{\partial z} = 0 \qquad (3)$$

If $\frac{\partial q_i}{\partial P_i}$, $\frac{\partial q_i}{\partial P_j}$ and $\frac{\partial P_i}{\partial z}$ are evaluated at the previous time step, Eq. (3) is a system of ordinary differential equations with unknown variables P_i. This can then be solved by the GEAR method for each distance position.

Using the method of characteristics, the time derivative of the adsorbed amount $\partial q_i/\partial t$ is first expressed by the following:

$$\frac{\partial q_i}{\partial t} = \left(\sum_{j=1}^{2} \frac{\partial q_i}{\partial P_j} \frac{dP_j}{dP_i} \right) \frac{\partial P_i}{\partial t} \qquad (4)$$

Then Eq. (2) becomes:

$$\left(\frac{\partial P_i}{\partial t}\right)_z + W_i \left(\frac{\partial P_i}{\partial z}\right)_t = 0 \qquad (5)$$

$$\text{where } W_i = \frac{u}{\varepsilon + RT\, \rho_B \sum_{j=1}^{2} \frac{\partial q_i}{\partial P_j} \frac{dP_j}{dP_i}} = \left(\frac{dz}{dt}\right)_{P_i} \qquad (6)$$

and W_i is the concentration wave velocity.

Coherence requires that the characteristic velocity for a given composition must be the same for all species:

$$W_1 = W_2 \qquad (7)$$

From Eqs. (6) and (7), we obtain:

$$\left(\frac{\partial q_2}{\partial P_1}\right)\left(\frac{dP_1}{dP_2}\right)^2 + \left(\frac{\partial q_2}{\partial P_2} - \frac{\partial q_1}{\partial P_1}\right) \frac{dP_1}{dP_2} - \frac{\partial q_1}{\partial P_2} = 0 \qquad (8)$$

The two roots for eq. (8) determine the two slopes of the characteristic lines in the hodograph plane (p_1 and P_2 diagram). For each composition (P_1, and P_2), there are two corresponding wavefront velocities that are calculated from Eq. (6). The final solutions of the original mass balance equation, Eq. (7), are obtained with the given boundary and initial conditions.

Results and Discussion

Because the adsorption isotherms of benzene and water vapor do not follow the Langmuir-type equation, an analytic solution for dP_1/dP_2 cannot be obtained. The construction of the hodograph must resort to numerical integration. The resulting characteristic lines are consequently not straight. Using the model given in Part I for predicting the adsorption equilibrium for mixtures of benzene and water vapor, the hodograph is constructed and shown in Fig. 1. Two families of characteristic lines can be identified: λ^+ corresponds to the higher wave velocity (for positive dP_1/dP_2); λ^- corresponds to the lower wave velocity (for negative dP_1/dP_2). In the low water vapor pressure range, the values of dP_1/dP_2 shift from positive to negative along the λ^+ curves, implying a slight enhancement of benzene adsorption due to the presence of water vapor. This synergistic sorption behavior, although yet to be experimentally proven, may be caused by the unfavorable isotherm of water vapor at low partial pressures. The hodograph constructed using Manes' model for adsorption equilibrium is shown in Fig. 2. A comparison of Figs. 1 and 2 shows that both methods exhibit similar behavior at high water vapor pressures. However, at low water vapor pressures, the hodograph based on Manes' model gives rise to straight vertical λ^- lines and straight horizontal λ^+ lines, implying no interference between the two adsorptives in this range.

Bed Profiles

With the aid of the hodograph, the concentration profiles at any given time in the bed can be readily calculated for any given feed and initial bed compositions by using the following procedure. Locate the feed and initial points on the hodograph, and find the intersection of λ^- curve that passes through the feed point and λ^+ curve that passes through the initial point. The intersection gives the plateau concentration plus two simple waves if the following two conditions are met: (1) wave velocity increases from feed point to the intersection point along the λ^-; and (2) wave velocity increases from the intersection point to the initial bed composition point along the λ^+ curve. Violation of either condition will result in a shock wave, the wave velocity of which is expressed by:

$$W_i = \frac{u}{\varepsilon + RT\rho_B \frac{\Delta q_i}{\Delta P_i}} \qquad (9)$$

where Δq_i and ΔP_i are the adsorbed phase and the gas phase, respectively, concentration differences between the plateau state and the feed state (or initial state) that are separated by the shock. From the coherence condition, Eq. (7), a new plateau concentration point can be found, which could be different from the intersection point due to the nonlinear characteristic curves in the hodograph plane.

As an illustrative example, we consider the S carbon bed initially equilibrated with 30 mmHg water vapor and 0.5 mmHg benzene vapor. The feed gas with an inlet velocity u=100 cm/s containes 28 mmHg water and 0.01 mmHg benzene and is at a total pressure of 1 atm and 30°C. The point of intersection is at (P_1 = 0.26mmHg P_2 = 26.65 mmHg). The wave velocity of this point along the λ^- characteristic is 2.62 x 10^{-3} cm/s, which is greater than the feed point velocity of 0.43 x 10^{-3} cm/s, and the wave velocity of intersection point along the λ^+ characteristic is 1.72 x 10^{-2} cm/s, which is less than the initial point velocity of 0.51 x 10^{-1} cm/s. Therefore, we can expect two simple waves and a plateau zone, as shown in Fig. 3 for t = 18 min in a 60 cm column.

The profiles calculated by directly solving Eq. (2) and the proposed thermodynamic model using finite difference method are also shown in Fig. 3 for comparison. The finite difference solution should approach the hodograph solution if a large number of cells is used for the column (40 cells are used in this calculation.). Although time-consuming, the numerical method has its advantages because for more complex situations, mass transfer, dispersion, heat transfer and varying velocity can all be readily incorporated in the model.

Water Condensation in the Bed During Adsorption

Because water vapor is the less strongly adsorbed component, a roll-over phenomenon for water vapor is expected in the case of adsorption, i.e., starting from a clean bed. Consider the example of an influent containing 24 mmHg H_2O and 1.0 mmHg C_6H_6 introduced into a clean column of carbon S, at u = 100 cm/s and 30°C. The bed profile predicted by the hodograph method for t = 100 min. is shown in Fig. 4a.

The plateau state has a water vapor concentration of 29.41 mmHg, which is higher than the influent water partial pressure, but lower than the water saturation pressure of 31.81 mmHg at 30°C. If the influent water partial pressure is increased from 24 to 29 mmHg, a water partial pressure greater than 31.81 mmHg at the plateau state is predicted by the hodograph method. This supersaturation will cause water vapor to condense in the bed. The maximum allowed water vapor pressure is 31.81 mmHg and the amount of the adsorbed-phase water (condensate) can be very substantial. Consequently, the velocity of the water concentration wavefront is reduced. To calculate the concentration profile, Eqs. (7) and (9) are used to calculate the benzene pressure and the amount of condensed water at the plateau state so the mass balance equation is still satisfied. The condensed water calculated for this example is 0.03 mole/g-carbon, or equivalent to approximately 1.3 times the micropore volume of the carbon. The benzene partial pressure in the plateau zone is 0.979 mmHg, near its feed partial pressure, as shown in Fig. 4b.

A comparison of Figs. 4a and 4b also shows that the bed coverage by the concentration wavefront is reduced when water condensation occurs. Furthermore, it is obvious that the occurrence of water condensation is not restricted to the initially clean bed situation. Friday and LeVan (20) first pointed out and discussed extensively the condensation of water vapor during hot purge of a dryer.

The long tail of the water concentration wavefront in Fig. 4 is the direct consequence of the unfavorable portion of the water isotherm. As the initial bed concentration is zero for both benzene and water vapor, the λ^+ characteristic that passes through the origin coincides with the abscissa, which is pure water vapor. Starting from the origin, the concentration wave velocity will decrease with the increase of water vapor pressure and then increase due to the S-shaped water isotherm.

As a result, a combination of shock wave and simple wave arises. The transition point, points S in Fig. 4, is determined by matching the shock wave velocity with the simple wave velocity in the λ^+ characteristic line, i.e., finding a point in the abscissa that has a λ^+-characteristic wave velocity with the same value as the shock velocity of the plateau state. This point, however, must be in the unfavorable portion of the water vapor isotherm so that the simple wave can exist. Accordingly, P_2 at point S is 20.72 mmHg in Fig. 4a, and is 20.81 mmHg in Fig. 4b. This behavior has not been encountered when both components have favorable adsorption isotherms. Similarly, in the case of water condensation, a long tail of water wavefront or a much earlier breakthrough of water is predicted as shown in Fig. 4b.

Acknowledgement

This work was supported by the Army Chemical Research and Development Center under Contract No. DAAA 15-85-C-0046.

LITERATURE CITED

1. Myers, A.L. and Prausnitz, J.M., AIChE J., II, 121 (1965).
2. Yang, R.T., "Gas Separation by Adsorption Processes", Butterworth, Boston, (1986), Chap. 3.
3. Grant, R.J. and Manes, M., Ind. Eng. Chem. Fundam., 5, 490 (1966).
4. Gregg, S.J. and Sing, K.S.W., "Adsorption, Surface Area and Porosity", Academic Press, NY (1982).
5. Dubinin, M.M., Carbon, 18, 355 (1980).
6. Dubinin, M.M., Carbon, 19, 402 (1981).
7. Barton, S.S., Evans, M.J.B., Holland, J. and Koresh, J.E., Carbon, 22, 265 (1984).
8. Okazaki, M., Tamon, H. and Taei, R., J. Chem. Eng. Japan, 11, 209 (1978).
9. Manes, M., in "Fundamentals of Adsorption", ed. A.L. Myers and G. Belfort, Engineering Foundation, New York (1983), p.335.
10. Dubinin, M.M., in "Chemistry and Physics of Carbon", Ed. P.L. Walker, Jr., Marcel Dekker, New York (1966), Vol. 2, p. 51.
11. Dubinin, M.M., in "Progress in Surface and Membrane Science", Ed. D.A. Cadenhead, J.F. Danielle, and M.D. Rosenberg, (1975), Vol. 9, p. 1.
12. Dubinin, M.M. and Zhukovskaya, E.G., Izv. Akad. Nauk. SSSR, Otd. Khim. Nauk., 670 (1957).
13. Rozwadowski, M., Siedlewski, J. and Wojsz, R., Carbon, 17, 411 (1979).
14. Rozwadowski, M. and Wojsz, R., Carbon, 19, 383 (1981).
15. Glueckauf, E., Disc. Faraday Soc., 7, 12 (1949).
16. Helfferich, F. and Klein, G., "Multicomponent Chromatography", Marcel Dekker, New York (1970).
17. Rhee, H., Aris, R. and Amundson, N.R., Phil. Trans. Roy. Soc., 267, 419 (1970).
18. Rhee, H. and Amundson, N.R., Chem. Eng. J. 1, 241 (1970).
19. Pan, C.Y. and Basmadjian, D., Chem. Eng. Sci., 26, 45 (1971).
20. Friday, D.K. and LeVan, M.D., AIChE J., 31, 1322 (1985).

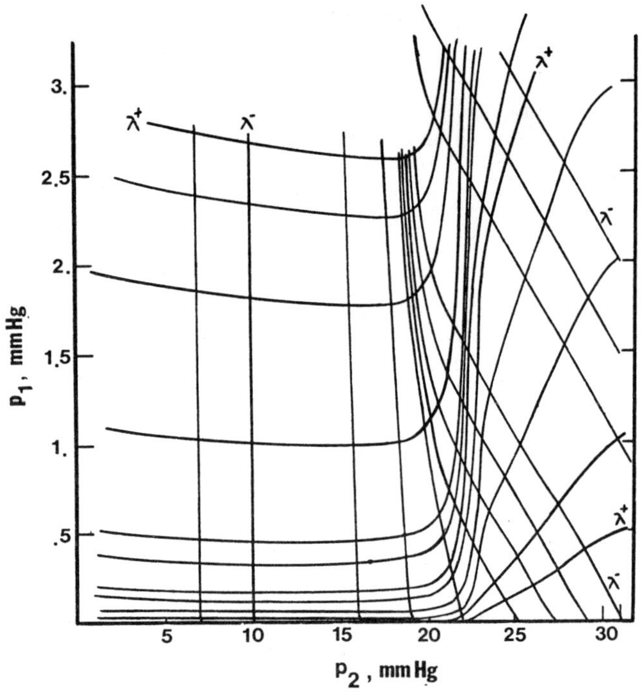

Figure 1. Hodograph plane of benzene and water vapor on Type S carbon using the proposed model for equilibrium mixture adsorption P_1 and P_2 and partial pressures of benzene and water vapor, respectively.

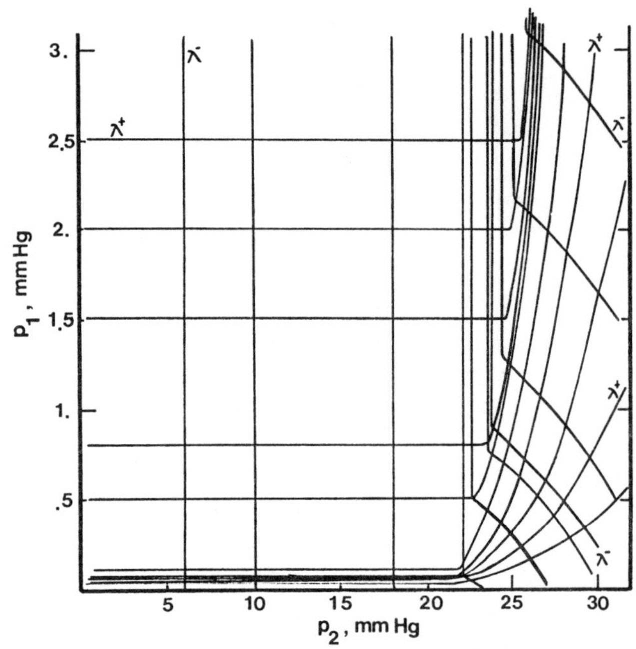

Figure 2. Hodograph plane of benzene (P_1) and water vapor (P_2) on Type S carbon using Manes' mode for adsorption equilibrium.

Table 1. Isotherm constants for pure-gas adsorption.

		Benzene	Acetone	Toluene	Methanol	Water
Molecular wt., MW		78	58	92	32	18
Liquid molar volume, V_L (cc/gmole)		89	73.5	106	42.5	18
Saturation pressure at 30°C, Ps (mmHg)		118.0	281.8	36.4	162.6	31.81
For Shirasagi S carbon	β	1	0.86	1.36	0.4	0.063
	$(\ln h_o)^2$	3.0	2.2	5.6	0.48	0
HG I-780 Carbon	β	1	0.85	---	0.36	0.059
	$(\ln h_o)^2$	---	2.2	---	0	0

v_o = 0.42 cc/g for Shirasagi S carbon

v_o = 0.28 cc/g for HGI-780 carbon

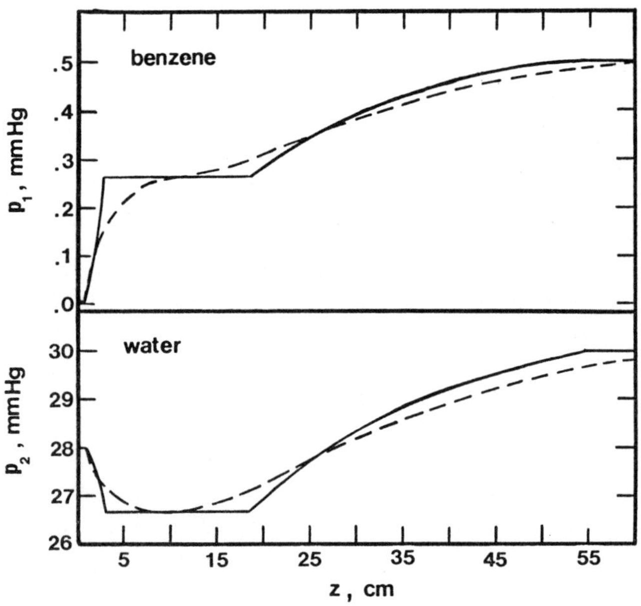

Figure 3. Bed concentration profiles of benzene and water vapor during desorption at t = 18 min, u = 100 cm/s, P = 1 atm, T = 30°C. The influent contains 28 mmHg and 0.01 mmHg benzene. The bed is presaturated with 30 mmHg water and 0.5 mmHg benzene ___ : hodograph method. - - - - : finite difference method.

Figure 4. (a) Bed concentration profiles of benzene (1) and water vapor (2) for a feed concentration P_1 = 1 mmHg, P_2 = 24 mmHg and initial bed concentration $P_1 = P_2 = 0$, using hodograph method. u = 100 cm/s, P = 1 atm, T = 30°C.

(b) Bed concentration profiles of benzene (1) and water vapor (2) showing the condensation of water when feed concentration is P_1 = 1 mmHg, P_2 = 29 mmHg using hodograph method. Other conditions are the same as in (a).

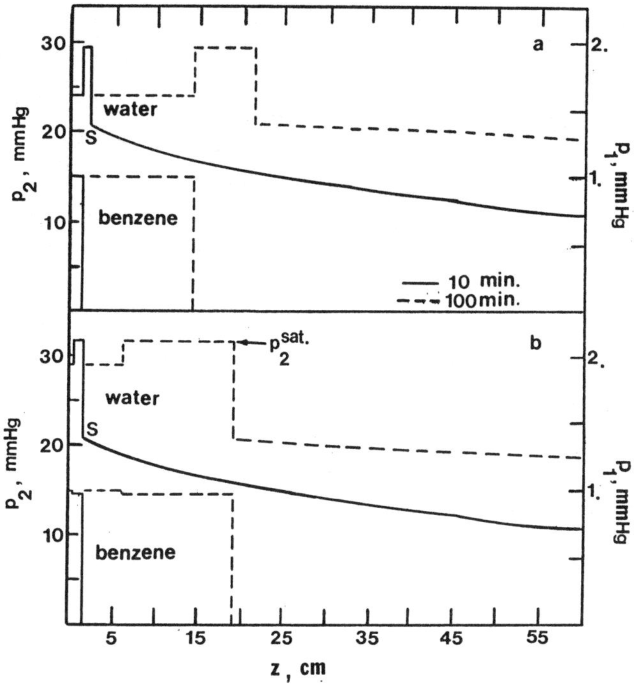

Table 2. Equilibrium adsorption of binary mixture of benzene (1) and water vapor (2) on Type S carbon at 30°C. Comparison of Experimental data (8) and model predictions.

		Experimental		This Model		Model in Ref. 8		IAS Theory		Manes' Model	
P_1,mmHg	P_2,mmHg	q_1,g/g	q_2,g/g	q_1,g/g	q_2,g/g	q_1,g/g	q_2,g/g	q_1,g/g	q_2,g/g	q_1,g/g	q_2,g/g
0.045	22.8	0.110	0.229	0.118	0.184	0.094	0.199	0.043	0.064	0.133	0.119
0.060	22.6	0.140	0.185	0.137	0.149	0.116	0.174	0.055	0.080	0.148	0.070
0.077	22.5	0.133	0.178	0.151	0.132	0.125	0.151	0.067	0.094	0.161	0.041
0.107	22.5	0.163	0.139	0.164	0.124	0.154	0.102	.084	0.114	0.178	0.022
0.312	23.1	0.238	0.079	0.184	0.184	0.204	0.099	0.140	0.175	0.228	0.038
0.489	22.4	0.269	0.031	0.235	0.075	0.237	0.019	0.184	0.207	0.261	0.000
0.081	28.1	0.068	0.274	0.086	0.303	0.109	0.276	0.027	0.054	0.113	0.267
0.086	27.7	0.076	0.253	0.090	0.295	0.113	0.262	0.030	0.059	0.119	0.256
0.163	28.1	0.108	0.221	0.107	0.280	0.152	0.227	0.047	0.087	0.144	0.232
0.212	28.1	0.142	0.196	0.116	0.270	0.168	0.207	0.057	0.101	0.157	0.217
0.262	28.7	0.175	0.160	0.120	0.270	0.185	0.194	0.061	0.107	0.162	0.218
0.313	28.1	0.173	0.150	0.132	0.253	0.196	0.176	0.074	0.124	0.177	0.195
0.432	27.9	0.212	0.147	0.148	0.234	0.225	0.134	0.093	0.146	0.195	0.171
0.601	27.9	0.220	0.121	0.165	0.216	0.250	0.106	0.113	0.166	0.213	0.151

Table 3. Adsorption of binary mixture of acetone (1) and water vapor (2) on Type S carbon at 30°C. Comparison of experimental data (8) with model predictions. (All amounts adsorbed, q, are in g/g).

		Experimental		This Model		Model in Ref.8		IAS Theory		Manes' Model	
P_1, mmHg	P_2, mmHg	q_1	q_2	q_1	q_2	q_1	q_2	q_1	q_2	q_1	q_2
0.150	25.6	0.043	0.323	0.059	0.293	0.031	0.377	0.033	0.055	0.071	0.267
0.716	25.6	0.092	0.259	0.103	0.245	0.089	0.296	0.099	0.136	0.138	0.183
1.002	24.4	0.098	0.223	0.128	0.203	0.106	0.276	0.135	0.166	0.166	0.121
1.337	23.1	0.129	0.225	0.154	0.159	0.122	0.248	0.173	0.191	0.194	0.053
2.831	24.5	0.195	0.135	0.178	0.153	0.206	0.145	0.206	0.219	0.226	0.052
6.123	23.8	0.236	0.089	0.232	0.093	0.251	0.091	0.273	0.255	0.271	0
9.284	22.2	0.276	0.059	0.279	0.028	0.281	0.038	0.328	0.278	0.304	0

Table 4. Adsorption of binary mixture of methanol (1) and water vapor (2) in Type S carbon at 30°C. Comparison of experimental data (8) with model predictions. (All amounts adsorbed, q, are in g/g).

		Experimental		This Model		Model in Ref.8		IAS Theory		Manes' Model	
P_1, mmHg	P_2, mmHg	q_1	q_2	q_1	q_2	q_1	q_2	q_1	q_2	q_1	q_2
4.420	28.6	0.023	0.338	0.046	0.345	0.055	0.356	0.147	0.124	0.039	0.353
4.493	25.5	0.053	0.267	0.056	0.295	0.073	0.321	0.183	0.142	0.054	0.286
4.635	24.3	0.069	0.278	0.065	0.265	0.080	0.301	0.200	0.150	0.064	0.248
6.524	26.1	0.112	0.242	0.071	0.288	0.102	0.289	0.210	0.159	0.079	0.268
6.848	23.3	0.119	0.219	0.095	0.216	0.099	0.274	0.251	0.176	0.107	0.168
28.431	20.4	0.166	0.160	0.275	0.116	0.204	0.150	0.427	0.252	0.316	0

Table 5. Adsorption of binary mixture on toluene (1) and water vapor (2) on Type S carbon at 30°C. Comparison of experimental data (8) with model predictions. (All amounts adsorbed, q, are in g/g.)

		Experimental		This Model		Model in Ref.8		IAS Theory		Manes' Model	
P_1, mmHg	P_2, mmHg	q_1	q_2	q_1	q_2	q_1	q_2	q_1	q_2	q_1	q_2
0.037	23.5	0.242	0.084	0.223	0.118	0.248	0.067	0.077	0.131	0.271	0
0.077	23.5	0.270	0.048	0.254	0.092	0.265	0.049	0.117	0.181	0.297	0
0.546	23.5	0.318	0.013	0.336	0.022	0.311	0.001	0.248	0.280	*	*
0.038	20.4	0.259	0.018	0.276	0.016	0.249	0	0.123	0.165	0.301	0
0.075	20.4	0.289	0.013	0.295	0.013	0.270	0	0.170	0.211	0.323	0
0.315	20.4	0.320	0.010	0.332	0.006	0.309	0	0.273	0.284	*	*

*Iteration did not converge.

Table 6. Adsorption of binary mixture of acetone (1) and water vapor (2) on Type HGI carbon at 30°C. Comparison of experimental data (8) and model predictions. (All amounts adsorbed, q, are in g/g.)

		Experimental		This Model		Model in Ref.8	
P_1,mmHg	P_2,mmHg	q_1	q_2	q_1	q_2	q_1	q_2
0.538	28.2	0.076	0.181	0.057	0.197	0.070	0.243
0.898	23.7	0.101	0.161	0.093	0.118	0.084	0.182
0.941	26.0	0.079	0.184	0.080	0.154	0.082	0.211
2.092	26.3	0.098	0.137	0.103	0.130	0.116	0.165
2.430	27.5	0.108	0.133	0.102	0.139	0.120	0.169
3.287	26.6	0.133	0.116	0.119	0.115	0.133	0.155
3.872	21.2	0.159	0.050	0.157	0.043	0.157	0.050
4.601	26.2	0.149	0.081	0.136	0.094	0.151	0.113
13.804	21.8	0.165	0.046	0.198	0.017	0.202	0.009

Table 7. Adsorption of binary mixture of methanol (1) and water vapor (2) on Type HGI carbon at 30°C. Comparison of experimental data (8) and model predictions. (All amounts adsorbed, q, are in g/g.)

		Experimental		This Model		Model in Ref.8	
P_1,mmHg	P_2,mmHg	q_1	q_2	q_1	q_2	q_1	q_2
4.962	27.2	0.053	0.193	0.024	0.228	0.046	0.266
12.023	21.3	0.109	0.117	0.09	0.090	0.082	0.158
22.313	22.2	0.122	0.126	0.126	0.076	0.117	0.136
23.798	20.1	0.130	0.096	0.147	0.039	0.127	0.081
28.089	18.8	0.142	0.079	0.166	0.020	0.143	0.063
33.780	21.2	0.153	0.105	0.168	0.036	0.163	0.077

INDEX

SYMPOSIUM SERIES

ADSORPTION

AEROSPACE

BIOENGINEERING

CRYOGENICS

CRYSTALLIZATION

DRAG REDUCTION

ENERGY

Conversion and Transfer

Nuclear Engineering

ENVIRONMENT

FLUIDIZATION

HISTORY OF CHEMICAL ENGINEERING

ION EXCHANGE

KINETICS

MINERALS

PETROCHEMICALS

PETROLEUM PROCESSING

PHASE EQUILIBRIA

PROCESS DYNAMICS

SEPARATION

SONICS

MISCELLANEOUS

MONOGRAPH SERIES